Hidden Risk in Occupational Electrical Safety

IEEE Press
445 Hoes Lane
Piscataway, NJ 08854

Hidden Risk in Occupational Electrical Safety

Through the Lens of System Safety

H. Landis "Lanny" Floyd
Founder and Principal Consultant
Electrical Safety Group Inc.
Elkton, Maryland, USA

Adjunct professor
The Advanced Safety Engineering and Management graduate engineering program
The University of Alabama at Birmingham
Birmingham, Alabama, USA

Staff electrical safety specialist
Chicago Electrical Trauma Rehabilitation Institute
Chicago, Illinois, USA

IEEE Press Series on Power and Energy Systems
Ganesh Kumar Venayagamoorthy, Series Editor

IEEE*Press*

WILEY

Published by John Wiley & Sons, Inc., Hoboken, New Jersey.

Library of Congress Cataloging-in-Publication Data Applied for

Hardback ISBN: 9781394257782

Cover Design: Wiley
Cover Images: © Kunakorn Rassadornyindee/Shutterstock, © kelvn/Shutterstock, © PINYO PROMPRASERT/Shutterstock, © Virrage Images/Shutterstock, © Andrewshots/Shutterstock

Set in 9.5/12.5pt STIXTwoText by Straive, Chennai, India

This book is dedicated to these individuals who illuminated the path in my journey exploring hidden danger in electrical safety:

Harry and Genevieve Floyd, Maryanna Hunt, Tom Ewers, O.J. Scott, Pam Floyd, Danny Liggett, Bill White, Don Wanner, Ralph Lee, Ed Troy, Bruce McClung, Bill Jordan, Andy Smith, Vince Flynn, Don Vardeman, Kim Eastwood, David Pace, Ray Crow, Christine Kent, Thad Brown, Mary Capelli-Schellpfeffer, Ray Jones, Bruce Cole, Ken Montgomery, Catherine Irwin, John Aeiker, Don Zipse, Dick Doughty, Deb Grubbe, Luiz Tomiyoshi, Dan Doan, Mike Doherty, Martha Bidez, and Anna Floyd.

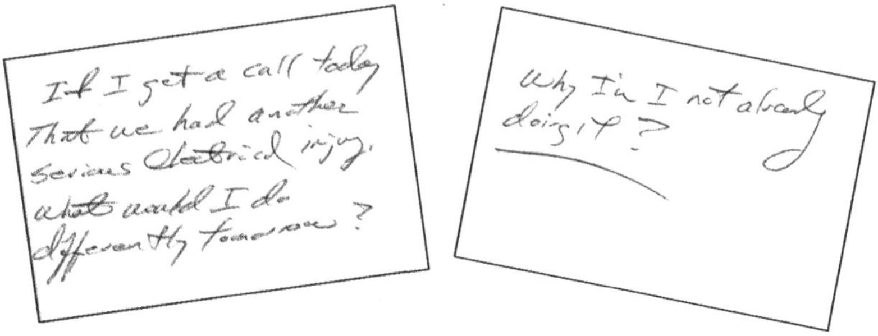

This is the front and back of a card I kept on my desk for the 20+ years that I led the DuPont Corporate Electrical Safety Team. The question and challenge served as a constant reminder to never stop exploring the residual risk that existed in our electrical safety management system. I've been retired from DuPont since 2014 and I still keep the card in my office. *Hidden Risk in Electrical Safety – Through the Lens of System Safety* is a record of my exploration, conducted with collaboration with my friends and colleagues in our continuous effort to pursue excellence in electrical safety.

H. Landis "Lanny" Floyd

We must challenge what we believe to be true.

Orville and Wilbur Wright
American aviation pioneers

This quote from Orville and Wilbur Wright reflects their frustration in attempting to build a machine that could fly. After several years of experiments and failure, they came to the conclusion that the formula to calculate wing lift, which had been in the scientific literature for more than 100 years, contained an error. They built a wind tunnel at their bicycle shop in Dayton, Ohio, to test their theory. The brothers confirmed their suspicion, determined the needed correction, and changed the world forever.

The lessons of the ordinary are everywhere. Truly profound and original insights are to be found only in studying the exemplary.

Warren Bennis
American Leadership Consultant

Contents

About the Author

H. Landis "Lanny" Floyd II, PE, CSP, CESCP, CUSP, CMRP, CRL, Life Fellow IEEE, has pushed the envelope in managing the risk of electrical injuries in the workplace for more than 50 years. His 45+ year career with DuPont focused on electrical safety in the design, construction, operation, and maintenance of DuPont facilities worldwide. In 2014, he joined the faculty of Advanced Safety Engineering and Management in the School of Engineering at the University of Alabama at Birmingham. His awards in recognition for his leadership and contributions in electrical safety include being a co-recipient of the IEEE Medal for Engineering Excellence in 2004.

Foreword

If you are considering reading this book, you obviously care about electrical safety and its special complexities. Your understanding of safety and safety culture could be especially important for your well-being and for your business success. You are wise to want to learn. Too many people really do not understand electrical safety, and sadly, some of them are no longer alive to share what they unfortunately learned in the last few seconds of their lives.

Why am I, a chemical engineer, encouraging you to read this book and to learn more about electrical safety? I'll give you the short version. It all started with my father, who was a high-voltage lineman for a very large utility in the US Midwest. I learned at six years of age that my dad could die at work. That was scary for me. As an only child, I became the son my dad did not have. Dad made sure that he taught me all the important things that a father would teach a son – how to throw a football with a perfect spiral, how to change the oil in the car, how to fish, how to drive a boat, how to shoot pool, and how to better understand that electricity is that unseen killer. He would say, "Honey, electricity is your friend. Most people do not pay attention to electricity, but it is always around, and you don't realize how important it is until you need it and you don't have it! Electricity makes your life much easier; however, you MUST respect it, because if you do not, it could kill you or hurt you severely. You cannot see electricity, you cannot hear electricity, you normally cannot smell electricity unless it is burning something else, you certainly cannot taste electricity, and if you ever touch electricity, it is very unforgiving and frequently deadly. Because it is so dangerous, there are a few rules: Rule #1: Always assume the wire is hot until you have tested it first to make sure there is no current. In other words, always TEST BEFORE TOUCH! Forget this rule, and we will be talking about you in the past tense! Rule #2: NEVER forget Rule #1!"

I think I was the only daughter sent out of the house after college with a pool table, an old boat, and an electric current tester! As a chemical engineer in the late 70s, I was fortunate enough to have my choice of job offers and took a job with the DuPont Company upon graduation. DuPont's lessons about safety sounded much like my father's advice, and I steadily learned about the engineering business. And that is where I met Lanny Floyd.

After having some experience in manufacturing facilities and building chemical plants overseas, I was approached by Lanny to work with him and the team of people he was pulling together. Our task was to improve the corporation's record around electrical fatalities and incidents in our manufacturing organization and construction projects. By then I had enough experience and perspectives that Lanny thought would be of value. I was surprised the team approached me to support their work; however, they did not know about my father's tutelage, and my personal passion for electrical safety and safety culture. It also gave me some early credibility with the team members who did not know me.

Electrical safety is much more than "Test Before Touch," but this rule is a good place to start. Sadly, though, too many people who work around electricity forget about this rule. In the context of this book, electrical safety and its supportive safety culture development becomes a constant and vigilant journey to acquire knowledge, procedures, rituals, and appropriate protections, all directed toward the goals of eliminating the terrible electrical incidents that maim and kill way too many.

Assuming you are interested in improving your Electrical Safety leadership, let's talk about what that journey could look like. It is a very satisfying journey; however, it is not easy and you cannot do it alone. This leadership journey, like most, is best done with others who feel the same passion for the subject. Electrical safety leadership includes learning about the technologies that have been developed to keep workers safer, about human behavior and what signs to watch out for, and about how to develop the safety systems needed to weave the first two items together into coherent protection for the worker and the asset.

So why should you care about Electrical Safety? It is all about preserving lives, limbs, and equipment, in that order. It is that simple and that complex.

Electrical Safety is and has been an important part of my life, and I would encourage you to consider it to be an important part of your life, too. Why? You and your loved ones live around electricity every day, and there are far too many electrical fatalities globally. According to 2025 OSHA data, there are over 2,000 lost time injuries annually in the United States due to exposure to hazardous electrical energy. One in 13 of those lost time injuries is fatal.

People my dad worked with died or received a severe electrical injury that caused the amputation of a foot in one case. People I worked with at DuPont died or were severely maimed. I will never forget the video I saw of one of my fellow employees explaining how he lost his arm and what that meant for him and his family. There was not a dry eye in the audience. Our children deserve to have their parents as knowledgeable and as safe as possible around one of the most common hazards each of us faces.

So, if you do not have an Electrical Safety Team where you work, please create one! From that first request from Lanny, I ended up becoming the first Corporate Operations Sponsor for the DuPont Corporate Electrical Safety Team, or CEST, as we called it. Our team established an Electrical Safety Month across the entire corporation each May. We had management support and were able to use corporate communications to not only reach every employee, but every contractor and every family member. Education is important for all, because everyone is exposed to the hazards of electricity.

Additionally, we also worked to rewrite our corporate electrical standard. We required our contractors to become more knowledgeable about our standard and contractually, they needed to follow it, as it was part of our procurement standard conditions.

The result of all of this was a significant reduction of electrical injuries over the years. I am very proud of what was accomplished by our electrical experts, technicians, engineers, and managers.

You will find this book to be an excellent roadmap for the key elements of Electrical Safety. May you read it with interest and purpose and thank you for being part of a continuing journey toward zero injuries in Electrical Safety.

Deborah L. Grubbe, PE, CEng, NAC, NAE
Chadds Ford, Pennsylvania

Preface

The hidden risk in occupational electrical safety lies in the residual risk that remains after compliance with regulations, codes, and standards. Assuming that compliance will achieve the desired degree of safety may result in unnecessary exposure to hazardous electrical energy. The lens of system safety is a mindset that can help illuminate the hidden risk and enable the implementation of additional risk controls that are reasonable, affordable, and effective. This book will help readers expose the hidden risk and push the envelope in reducing mishaps in electrical systems.

For more than 50 years, I have explored advancements in technology, safe work practices, management systems, and human and organizational performance as they relate to occupational electrical safety. During this journey, I have seen annual electrocution fatalities decrease by more than 80% in the United States. However, the downward trend has flattened for more than a decade. Other industrialized countries have demonstrated significantly better results in reducing electrical fatalities, which suggests that a significant reduction in the United States is possible.

My journey in electrical safety began when I was 11 years old, when my father enrolled in an electrician apprenticeship program with the DuPont Co. Our house was soon filled with tools of the trade and his friends in the program who gathered on the weekends to study. At age 19, I began an electrical engineering internship at the DuPont plant where my father worked. One month after I started my internship, the supervisor of the electrical engineering group called us together to share the news that a plant supervisor had suffered a fatal injury by contacting a utility 115,000 V overhead transmission line. I remember the senior engineers asking, "How could he have contacted a transmission line?" I did not realize it at that time, but that event shaped my career. I began collecting articles on electrical injuries and fatalities. After completing my engineering degree, I accepted a full-time position as an electric power engineer at the DuPont plant where I had interned. Electrical safety was at the forefront of my job, as I supported the operation and maintenance of the electric power system.

The 1979 accident at the Three Mile Island nuclear power station changed how I thought about preventing electrical incidents, and I began applying lessons from nuclear safety to my work in electrical safety. The accident led me to study system safety, the powerful safety management methodology developed during World War II to control the risk of emerging technologies with potentially catastrophic consequences of magnitudes never seen before. But in 1982, my confidence in electrical safety was shaken. I was with the company's electrical safety expert on a maintenance inspection at one of our plants. He was about 20 feet from me when he slipped and contacted an unguarded, energized 480 V conductor. An electrician and I performed CPR until the paramedics arrived. He died several days later. He was the most knowledgeable, competent, and passionate person in electrical safety that I knew. The question repeated throughout the company was, "How could this have happened to him?" Then, in 1984, my father's career as an electrician ended in an arc flash event. A short time later, I met with my DuPont manager for my annual performance review when I told him I had just one career goal: to make a difference in changing the electrical safety culture to prevent serious electrical injuries. Without hesitation, he responded, "How can I help?" His help came in the form of supporting me to get involved in IEEE and other professional organizations that had the potential to shape the future of electrical safety.

Throughout the 19th and 20th centuries, DuPont was known as a pace-setting leader in occupational and process safety. However, in the mid-1980s, several of my colleagues and I became concerned that the frequency of injuries from exposure to electrical hazards was an outlier in the company's overall safety performance. We collected data to support our concern. In 1989, top management made a highly visible commitment to reducing the risk of injuries from electrical hazards to employees and contractors. Management established goals for sustainable improvement, provided financial support, and empowered dedicated people to reduce the likelihood of electrical incidents. We began crafting a focused application of DuPont safety management expertise in highly hazardous processes to the unique hazards of electrical energy. We called this *Electrical Safety Management*, and its implementation resulted in sustainable breakthrough improvement in electrical safety performance in DuPont operations, as shown in Figure 1. *Electrical Safety Management* integrates electrical hazard identification and injury risk control measures into an overarching safety management system.

For more than 30 years of my 45+ year career with DuPont, I helped lead the work to change the electrical safety culture in DuPont. I had the privilege of working with influential colleagues who shared the vision that we could change the electrical safety culture. And we did. Figure 1 shows the breakthrough, sustainable performance achieved in the early 1990s. In the first 25 years of my career, there were 12 electrocution fatalities in our facilities. In my last 20 years, there were zero.

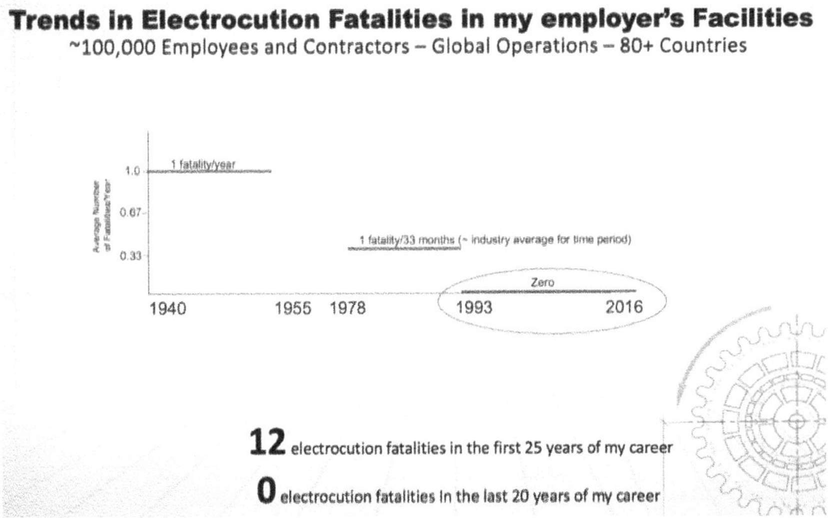

Figure 1 Trends in electrocution fatalities in my employer's facilities.

Since the passage of the US Occupational Safety and Health Act in 1970 and into the first three decades of the 21st century, occupational electrical safety in most organizations in the United States has centered on compliance with OSHA electrical regulations and consensus standards, such as NFPA 70E *Standard for Electrical Safety in the Workplace*. Compliance with electrical regulations and standards is part of the solution to reduce the risks of injuries and fatalities. There is an opportunity to change the electrical safety culture and look beyond compliance. Exploring electrical safety through the lens of system safety involves understanding how hazard identification, risk assessment, and risk reduction can supplement the results in injury prevention afforded by a compliance-based approach.

In Figure 2, the white space around the electrical safety regulations and standards is a topic in advanced safety engineering and management not addressed or discussed in detail in electrical safety regulations, codes, and standards prominent in the United States. From my experience in leading cultural change to enable sustainable improvement in preventing life-changing and fatal injuries from electrical hazards, continual learning on the topics in the white space is critical to going beyond compliance and integrating electrical injury risk reduction into a safety management system. You may find hidden risks in these topics that exist for your organization.

So, what have I learned during this 50-year journey? I have had the privilege of collaborating with hundreds of people to effectively advance the electrical safety culture to prevent injuries and fatalities. The potential for exposure to hazardous

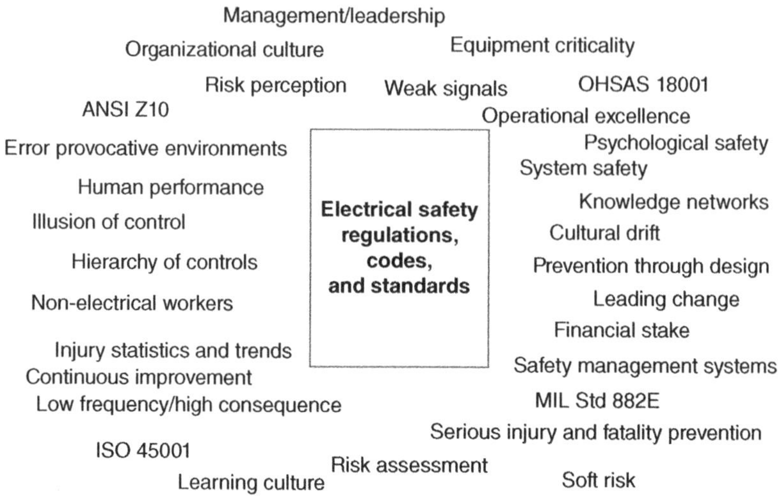

Figure 2 Opportunities to find hidden risk.

electrical energy has increased due to the growth in applying electrical technologies for energy, control, and communications in industrial applications. The evolution in occupational electrical safety regulations and standards in the United States has significantly reduced occupational electrical injuries. However, the approach taken in the United States has created a widely held paradigm that compliance alone is sufficient. Other countries have demonstrated that compliance combined with advanced safety concepts to more effectively identify hazards, assess risks, and evaluate risk control effectiveness can impact further reductions in injuries and fatalities from exposure to electrical hazards.

Significant improvement in preventing occupational electrical injuries and fatalities will require a culture shift in how we think about controlling exposure to hazardous electrical energy. Compliance-based programs must shift from compliance only to incorporate risk management and other concepts derived from system safety. Electrical safety leaders will need to commit to continual learning of the most effective approaches to safety management. Adding risk management components to compliance involves an in-depth understanding of hazard identification, risk assessment, and risk reduction to supplement the results in injury prevention afforded by a compliance-based approach. I hope this book is an aid in accelerating this change.

The journey continues…

July 7, 2025 *H. Landis "Lanny" Floyd*

Acknowledgments

I have had the privilege of collaborating with hundreds of people to advance the electrical safety culture to more effectively reduce exposure to hazardous energy and prevent injuries and fatalities. Regrettably I cannot list every individual. However, I am greatly indebted to colleagues who shared their wisdom, encouragement, experiences, and passion to advance the practice of electrical safety, beginning in 1984 with our work to create *IEEE Standard 902, Guide to Operation Maintenance and Safety of Industrial and Commercial Power Systems;* the IEEE Petroleum and Chemical Industry Safety Subcommittee chartered in 1991 and provided the environment to accelerate progress in advancing solutions to arc flash exposure and to launch the annual Electrical Safety Workshop; members of the DuPont Corporate Electrical Safety Team, which I worked to establish in 1989 and chaired from 1992 to 2014; the DuPont managers who supported and encouraged my career and actively championed excellence in electrical safety; the leadership team for the IEEE IAS Electrical Safety Workshop established in 1991 and led to establishing the IEEE IAS Electrical Safety Committee in 2012; the faculty and students in the Advanced Safety Engineering and Management graduate engineering program at the University of Alabama at Birmingham; and to the co-authors of my published papers.

I want to especially express my appreciation to several individuals.

Danny Ligget, with whom I've collaborated since 1976, and who along with Kim Eastwood and David Pace seized the opportunity to connect stakeholders to change and advance the electrical safety culture; Anna Floyd, PhD, for her long time interest in electrical safety and linking her background in social psychology, risky behaviors, and risk perception to develop Chapters 7, 11, and 13; and Pam Floyd, who has been my supporter, confidant, and advisor throughout my journey.

Introduction

The future has already arrived. It's just not evenly distributed.

William Gibson
American futurist and science fiction author

I have a confession. There is nothing new in this book that has not already been published or presented elsewhere. The content of *The Hidden Risk in Occupational Electrical Safety* is a synthesis of hundreds of papers, articles, conference presentations, tutorials, and webinars that I've authored over the past 40+ years. Each chapter is centered on central themes from my previously published work. So what is the value of this book to you? The subtitle, *Through the Lens of System Safety,* is a label I recently applied to the mindset I've used in exploring the nooks and crannies of electrical safety-related management systems, safe work practices, and technology to identify opportunities for further reducing the risks associated with exposure to hazardous electrical energy. Many potential readers may not be aware of or have access to my past publications and presentations. After more than 40 years of championing opportunities to advance the practice of electrical safety through my papers and presentations, this book provides the tool, *The Lens of System Safety*, to show these opportunities in a cohesive and connected way.

Imagine driving on an isolated rural road at night, with light coming only from a full moon. The low level of lighting is sufficient to navigate the roadway. Switching on the headlights changes the details you can see, giving you more ability to see hazards and avoid risks. Similarly, the mindset of system safety can help expand how you think about simply complying with regulations, codes, and standards. Gaps and opportunities in management systems, safe work practices, and technology are illuminated, which can lead to applying additional measures or controls to further reduce acceptable risk.

The **Preface** describes some of the experiences and relationships from my personal and professional life that had a particular impact on shaping my view on how I could make a difference in preventing serious injuries and fatalities caused by electrical energy.

In **Chapter 1, "The Lens of System Safety,"** I emphasize both the importance and limitations of an electrical safety culture based on simply complying with the Occupational Safety and Health Act (OSHA) regulations, which first appeared in 1970, followed by compliance resources provided by the National Fire Protection Association (NFPA), IEEE, and other standards developers. I discuss a more comprehensive solution based on proven concepts derived from system safety. System safety has its roots in the efforts to control the risk of catastrophic accidents in nuclear weapons and advanced aviation that evolved during World War II. The lens of system safety is a mindset framed by the most rigorous safety and risk management methodology embedded in system safety. The lens of system safety is the foundation for thinking differently, enabling me to understand how to embrace a risk-based approach that complements compliance.

On March 28, 1979, an accident began unfolding at the Three Mile Island nuclear generating station in Harrisburg, PA. **Chapter 2, "What We Can Learn from a Nuclear Accident,"** tells a story about how a nuclear accident, having the potential of being the most catastrophic man-made disaster in human history, introduced me to the power of system safety. This chapter discusses a key learning from *The Report of the President's Commission on the Accident at Three Mile Island.* That key learning helped me understand that compliance with regulations and standards was essential in the journey, but was not the destination, in my work to advance electrical safety.

Electrical safety regulations, codes, and standards provide minimum requirements for preventing electrical injuries. In **Chapter 3, "Limitations of a Compliance Culture,"** residual risk is described as the amount of risk associated with a task or process that remains after inherent risks have been reduced by an acceptable risk control methodology, i.e., compliance. What if intrinsic risks are not completely identified? How do you determine if the residual risk has been reduced to a level acceptable to both the worker and the organization? This chapter explores these and other questions to provide insight into methods for searching and exposing opportunities for continual improvement, thereby reducing worker exposure to hazardous electrical energy. Areas of potential hidden danger discussed include management/leadership, facilities design, safe work practices, incident investigations, procurement, the included workforce, and outsourcing.

Chapter 4, "A Case History in Breakthrough Performance," discusses the demonstrated breakthrough and sustainable results from the electrical safety improvement strategy implemented by DuPont in its global operations. DuPont was a pace-setting leader in process and occupational safety throughout the 19th and 20th centuries; however, in the mid-1980s, I was among a handful of people concerned that electrical injuries were an unrecognized outlier in the company's safety performance. We utilized data from a workforce of over 100,000 spanning nearly 20 years to help top leadership understand that accepted exposure to hazardous electrical energy was a phenomenon of very low frequency but extraordinarily high consequence. The low frequency of electrical injuries had caused a blind spot in safety management. However, our data showed that approximately 8% of the lost time injuries (LTIs) from hazardous electrical energy were fatal. Although nonfatal electrical injuries had almost no impact on calculating LTI metrics, the number of fatalities was a significant contributor to the number of deaths in DuPont's experience. Two of my published papers serve as bookends for this chapter. The first paper, published in 1993, outlined our strategy for changing the electrical safety culture in DuPont. The second paper, published in 2013, discusses 20 years of experience in continually improving the implementation of the strategy.

We live in an electrical world, and all workers are exposed to some degree of electrical hazards. For office workers, exposure may be limited to appliances, cords, outlets, and power strips. For construction laborers, exposure includes the use of extension cords, power tools, and overhead and underground power lines. Overhead lines pose a concern, as unintentional contact with mobile equipment, scaffolds, ladders, and conductive materials is a leading cause of fatalities among construction workers. **Chapter 5, "Our Electrical World,"** poses a key question: Is your electrical safety program applied to workers whose job expectations include working on or near energized circuits and equipment, or is it used for all workers who may be exposed to electrical hazards? The chapter discusses how the electrical safety program should encompass all workers, with the program tailored to the specific hazards and risks in various work environments.

Modern industry and commerce are nearly entirely dependent on electrical technologies for energy, control, data management, and communications. While a worker's mishap in these critical systems may result in injury to the worker, with much greater frequency, an accident or incident also has the consequence of significant disruption to the enterprise's operations. **Chapter 6, "Collateral Injuries, Consequences, and Benefits: The Underlying Power of an Effective Electrical Safety Program,"** discusses the broad implications of incidents in electrical systems critical to business and commerce. Ensuring that top management understands the total value of continual improvement in preventing electrical mishaps can help provide sustained value and support to justify measures and investments in electrical safety programs, thereby preventing injury to workers.

Recognizing personal vulnerability to risk is an essential step in establishing objectives for the design of electrical systems and in improving workers' safety-related behaviors to reduce injury rates. Studies in the fields of cognitive, social, and health psychology suggest that risk perception catalyzes changing behaviors, including those related to safety. **Chapter 7, "The Value of Vulnerability,"** combines research findings in psychology with practical engineering experience regarding high-risk hazards that have potential for fatal and disabling injury and shows how vulnerability can impact the effectiveness of safety training for engineers who design electrical systems, supervision of workers potentially exposed to electrical hazards, and workers at risk for serious electrical injury.

The scopes of OSHA electrical safety regulations and NFPA 70E *Standard for Electrical Safety in the Workplace* address safe work practices and personal protective equipment (PPE), which are *part of a comprehensive solution* for managing exposure to hazardous electrical energy. Suppose this limitation is not understood when applying these standards. In that case, an organization seeking to reduce or eliminate electrical incidents and injuries in its operations may not fully realize the potential of achieving this goal. **Chapter 8, "Safety Management Systems,"** discusses the benefits of implementing an electrical safety program that incorporates the requirements of electrical safety regulations and standards within the management framework of Occupational Health and Safety Management Systems (OHSMS) standards, including ANSI Z10 and ISO 45001. Effective implementation and incorporation into an OHSMS serve to ensure continual improvement and sustained support from top management.

In recognition that the United States lags other countries in applying engineering design solutions to reduce the risk of injury, the National Institute for Occupational Safety and Health (NIOSH) launched the Prevention through Design (PtD) initiative in 2007. **Chapter 9, "Prevention through Design,"** discusses innovation in policies, regulations, and standards that support the application of design techniques, reducing the risk of injury throughout the full life cycle of installations, including construction, commissioning, operations, maintenance, renewal, and demolition.

Chapter 10, "The Intersection of Reliability and Safety," presents strategies to optimize the integration of an organization's electrical safety program into business management systems for critical equipment maintenance. The chapter describes methods for identifying risks, establishing priorities, allocating resources, and assessing the quality of efforts to maintain the designed functionality of equipment critical to electrical safety. It further examines the relationship between the electrical safety program and the maintenance and reliability program. It suggests ways to enhance the synergy and integration of these two business management systems.

In the context of the hierarchy of risk controls, warnings, administrative controls, and PPE are sometimes referred to as lower order controls. In practice, the effectiveness of these controls can be overestimated due to a misunderstanding of the human factors that affect error likelihood in their implementation.

Chapter 11, "Residual Risk and the Psychology of Lower Order Controls," explores one human factor – the limitation of attention – on residual risk inherent in lower order controls. The limitation of attention has implications for identifying hazards, assessing risk, taking appropriate actions based on warnings, executing safe work practices, and selecting and using PPE properly. Raising awareness and improved understanding of these topics can enable more effective risk assessment, quantification of residual risk, and more effective selection and combination of risk controls to achieve risk as low as reasonably practicable (ALARP).

Chapter 12, "Measure What You Need to Manage," explores opportunities to complement traditional methods of measuring safety performance with leading indicators, providing a balanced scorecard of both lagging and leading indicators to drive continual improvement in reducing the risk of electrical injuries. Lost-time electrical injuries are among the rarest types of injuries in the workplace, accounting for less than 0.2% of all workplace injuries in the United States. Yet, contact with electrical energy is among the top 10 causes of occupational fatalities. These two attributes – very low frequency and very severe consequences – create a serious challenge in measuring the effectiveness of electrical safety programs. The traditional measurement of injury rates, a lagging indicator, may blind an organization to future injury potential. The low frequency of electrical injuries may result in an organization having insufficient data points to make the results statistically valid.

Chapter 13, "End Game," examines cultural drift with respect to occupational electrical safety. It poses the question of how the reduction in occupational electrocution fatalities over the past 40 years could present new challenges in maintaining electrical safety vigilance today, including the impact on electrical injury-related risk perceptions of workers, supervision, and management. The chapter discusses how low-frequency/high-consequence events pose unique challenges for safety management and provides strategies to counteract cultural drift.

Credible data is essential in telling the story of electrical safety. Not everything that comes across your desk or computer screen may be factual, even if created with good intent. The **appendix** provides references to credible sources. There may be more sources of credible, detailed data on electrical injuries and fatalities. Still, three notable organizations are the NFPA Fire Protection Research Foundation, CPWR – The Center for Construction Research and Training, and the Electrical Safety Foundation International. These three organizations periodically analyze data from the US Bureau of Labor Statistics (BLS) and other credible sources, publishing reports that are useful in identifying opportunities to improve electrical safety efforts. Their reports can be downloaded free of charge.

1

The Lens of System Safety

Your occupational electrical safety program may need a reset. This book can help you discover hidden risks that can be reasonably and feasibly reduced using techniques that have demonstrated results in preventing serious injuries, fatalities, and other catastrophic consequences.

Since the passage of the Occupational Safety and Health Act (OSHA) in 1970, the discussion of occupational safety in the United States has focused mainly on compliance with safe work practices in OSHA regulations and industry consensus standards. The regulations and standards have and will continue to help prevent serious injuries and fatalities. However, regulations and standards provide minimum requirements. Compliance with minimum requirements may create an illusion for management and workers that risk is reduced to as low as reasonably practicable (ALARP). A gap may exist between risk reduction achieved from compliance with minimum requirements versus risk reduction achieved by other means, or the difference between the accepted risk and the acceptable risk. ALARP expectations will vary from organization to organization and from one individual to another. What is considered acceptable risk will change over time, impacted by various forces, including social consciousness, competitive environments, and leadership goals.

OSHA regulations and industry standards relevant to electrical safety provide requirements and guidance to control exposure during work activity. However, electrical safety does not start during the installation, operation, and maintenance of electrical systems in the workplace. Decisions made during the design, specification, and procurement of equipment, components, and tools can have a positive or negative contribution to exposure to hazardous electrical energy. These decisions occur long before the workplace exists and have an impact on the exposure for the life of the facility or equipment.

Hidden Risk in Occupational Electrical Safety: Through the Lens of System Safety,
First Edition. H. Landis "Lanny" Floyd.

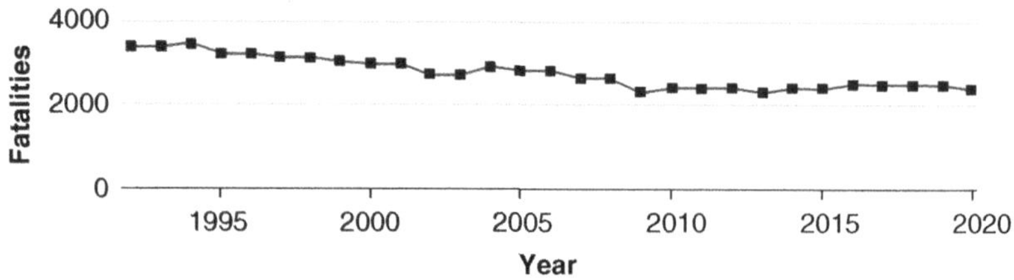

Figure 1.1 The trend in occupational fatalities from all causes in the United States, 1990–2020.

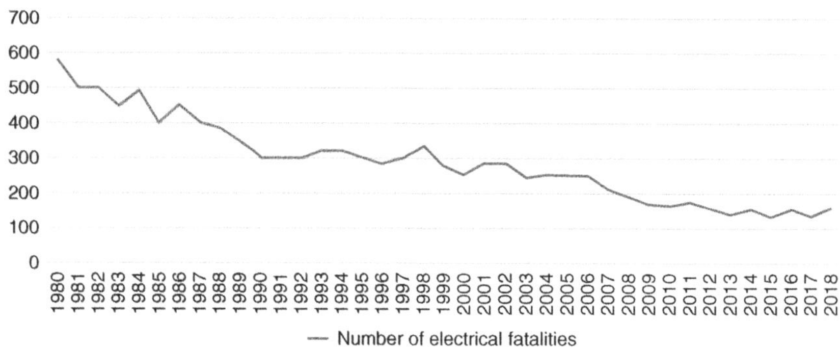

Figure 1.2 The trend in occupational fatalities from exposure to electrical hazards in the United States, 1980–2018.

Thought leaders in safety management, including Peterson, Krause, and Manuele, have expressed concern about the trend in occupational fatalities in the United States. During this period, the downward trend in the number of fatal injuries from all causes and the subset of fatalities from exposure to hazardous electrical energy has flattened (Figures 1.1 and 1.2).

In a compliance-based culture, organizations adhere to standards and regulations, which can lead to a "check-the-box" approach to comply with the minimum requirements provided by standards and regulations. A risk-based approach helps identify risks that remain even after achieving compliance. The residual risks can then be assessed to determine if additional controls are warranted.

Organizations can be blinded by the data and fail to see the need to go beyond minimal compliance. Lost-time injuries from exposure to hazardous electrical energy are relatively rare in the context of all lost-time occupational injuries. During the third decade of the 21st century, less than 0.2% of all occupational lost-time injuries in the United States are attributed to injuries from exposure to hazardous electrical energy. The very low frequency of electrical injuries in an individual's or organization's experience can create an illusion of the effectiveness of compliance-based risk control in two ways:

1) It can create a false perception for management that hazards have been identified and risk controls are adequate.
2) It can be challenging for workers to stay vigilant and maintain a perception of risk in the face of a low probability of occurrence.

Andersen and Denkle noted that there is a wide range of severity associated with occupational lost-time injuries. Their research showed that 1 in 250 lost-time injuries from all causes was fatal. However, Cawley and Brenner reported that 1 in 13 lost-time injuries from exposure to hazardous electrical energy is fatal. The low frequency of electrical injuries can lead to the electrical hazards being accepted, the risk of injury being underestimated, and the effectiveness of mitigation controls being overrated (Table 1.1).

A risk-based approach to managing occupational safety reaches beyond compliance with the minimum requirements of regulations and standards. In a risk-based approach, compliance with standards and regulations is a means, but not the end, of hazard identification and applying risk control measures. Even after achieving compliance, residual risk may be further reduced using hazard identification and risk control evaluation tools. The most comprehensive safety and risk management methodology is system safety; however, its use has historically been limited to the never-before-seen catastrophic safety risks that

Table 1.1 Comparison of select non-fatal occupational injuries in the United States 2010.

Type of Non-Fatal Injury	Number of Injuries
Total	1, 191, 100
Sprains, strains, tears	474, 000
Musculoskeletal disorders	346, 300
Falls on the same level	182, 400
Struck by object	138, 530
Falls to lower level	73, 520
Assault/violent act by person	40, 310
Highway accidents	36, 460
Assault/violent act by animal	7, 160
Fires and explosions	3, 000
Electrical shock and burn	**1,890 (0.16%)**

Adapted from U.S. BLS Economic News Release, 2010.

emerged in the 20th century, such as nuclear weapons and commercial aviation. What if electrical safety professionals were more familiar with concepts derived from system safety? System safety has its roots in the 1940s, with the rapidly expanding military and commercial aviation industries and nuclear weapons technology. Designers, manufacturers, and pilots were pushing the envelope in technology. As airplanes became more sophisticated, the cost of mishaps escalated. The aviation industry recognized that analyzing mishaps after the fact was becoming unacceptable in terms of human safety and financial loss. In his paper, "Using System Safety Techniques to Perform Hazard Analysis," Kolak noted, "The costs associated with damaging expensive fighter jets and the development of nuclear energy (where a single system failure was unacceptable) contributed to the concept that hazards must be anticipated and controlled before even a single loss occurs." These emerging technologies, operating at the edge of engineering capability, using unproven technology, required conservatism to protect against unknown mishap risks and allowed for novel solutions.

System safety evolved to address exotic technologies with never-before-seen catastrophic consequences. It was not intended to be applied in managing the risks associated with the more common hazards in the workplace. Table 1.2 lists more than 20 risk assessment tools and techniques used in system safety. The costs and resources needed to apply system safety to all hazards and risks would be unreasonable and impractical. However, concepts learned from system safety were instrumental in developing regulations and standards for highly hazardous

Table 1.2 Hazard identification and risk analysis tools used in system safety.

Preliminary hazard analysis	Sneak circuit analysis
What if analysis	Layers of control analysis
Subsystem hazard analysis	Petri net analysis
System hazard analysis	Markov analysis
Operating and support hazard analysis	Barrier analysis
Health hazard analysis	Bent pin analysis
Layers of protection analysis	Hazard and operability analysis
Fault tree analysis	Cross-consequence analysis
Event tree analysis	Common cause failure analysis
Failure mode and effect hazard analysis	Management oversight risk tree analysis
Fault hazard analysis	Software safety assessment
Functional hazard analysis	Bow tie analysis

chemical processes in the early 1990s, including OSHA 1910.119 Process Safety Management and the Environmental Protection Agency's Risk Management Program. The demonstrated results of system safety influenced the development of ISO 45001 in 2018 and also the 2019 revision to ANSI/ASSP Z10. The critical concept learned from decades of experience with system safety is that it can provide the "lens" to illuminate and identify risks that remain after compliance with regulations and standards is achieved.

System safety engineering evolved as a way to minimize the consequences and likelihood of mishaps by eliminating causes attributed to equipment design, equipment reliability, and human error. In 1969, the U.S. Department of Defense published MIL-STD 882, *Standard Practice for System Safety*. Initially published for managing risks associated with the first-generation nuclear ICBM deployment, this standard has provided the basis for managing the safety of complex systems such as military aviation, nuclear weapons, nuclear-powered ships and submarines, as well as less complex systems such as facility operations. The standard has influenced the application of safety system engineering to non-military systems, including commercial nuclear power generation, commercial aviation, space travel, and highly hazardous petrochemical operations. The following definitions from MIL-STD 882E, *Standard Practice for System Safety*, provide an introduction to some of the fundamental concepts of systems safety engineering:

> Mishap: An unplanned event or series of events resulting in death, injury, occupational illness, damage to or loss of equipment or property, or damage to the environment.
>
> System: An integrated composite of people, products, and processes that provide the capability to satisfy a stated need or objective.
>
> System safety: The application of engineering and management principles, criteria, and techniques to achieve an acceptable mishap risk, within the constraints of operational effectiveness and suitability, time, and costs, throughout all phases of the system life cycle.

System safety engineering concepts include hazard identification and analysis, application of a hierarchy of hazard control measures, and integration of management, engineering, procurement, and operations functions in managing safety objectives. These concepts are the foundation of occupational safety and health management systems standards that emerged since the 1990s, including OHSAS 18001 *Occupational Safety and Health Management Systems – Requirements*, ANSI Z10 *Occupational Health and Safety Management Systems*, CSA Z1000 *Occupational Safety and Health Management*, and ISO 45001 *Occupational Safety and Health Management Systems.*

Demonstrated Performance

Research by Mendleoff and Staatsky shows that the countries in the European Union (EU) have a significantly lower occupational fatality rate than the United States, with the U.S. fatality rate being ~4× that of the United Kingdom. Although there are cultural, social, educational, and political differences that may contribute to the lower fatality rates, thought leaders in safety point to the broad application of risk assessment processes as a significant difference between the United Kingdom and the United States. Fred Manuele noted that risk-based culture is the cornerstone of the EU approach to occupational safety management. The American Society of Safety Professionals, OSHA, and other safety institutions have promoted a risk-based approach as critical to further improvement in injury and fatality reduction.

For over 75 years, system safety has delivered extraordinary results in reducing risks of 20th-century technological advancements that have the potential for catastrophic mishaps, including nuclear energy, advanced weapons systems, aviation, and manned space exploration.

Exploring electrical safety through the lens of system safety involves understanding hazard identification, risk assessment, and risk reduction to supplement the results in injury prevention afforded by a compliance-based approach. To create a lens of system safety and enable this exploration, individuals and organizations must recognize that simple compliance with the minimum requirements of regulations and industry standards always has residual risk that may not be ALARP. Electrical safety leaders should commit to continual learning on system safety and risk-based safety management.

After achieving compliance with safe work practices, applying risk management techniques includes exploring answers to questions, such as "How can the risk controls fail? What are the consequences if one or more controls fail? Are the potential results acceptable?" These questions are fundamental to system safety. In the context of possible exposure to hazardous electrical energy, asking and answering these questions can help see beyond compliance with minimum requirements to how hidden hazards can be identified and how residual risk can be further reduced. Electrical safety leaders can explore these questions throughout the life cycle of a facility or system, including initial design, equipment specification and procurement, task planning, job safety analysis, and incident investigations.

Demonstrating Results in Moving Beyond Compliance

In 1989, DuPont, a global Fortune 500 company made a highly visible commitment to reduce the likelihood and severity of injuries to employees and contractors from

Table 1.3 Elements of the strategy described in the paper "Creating a Continuous Improvement Environment for Electrical Safety."

- Understand the business consequences of electrical incidents
- Engage all employees
- Stimulate near-miss reporting
- Apply quality improvement model – Plan Do Check Act
- Build networks
- Challenge accepted practices
- Improve collaboration among management, electrical experts, and safety professionals
- Use standards as tools
- Promote prevention by design
- Address life cycle: design, construct, operate, maintain, dismantle

electrical hazards. Goals for sustainable improvement were established, financial support provided, and dedicated people empowered to reduce the likelihood of electrical incidents, injuries, and fatalities, with the intent to accomplish a step change in electrical safety performance, as was done in the mid-1950s. At that time, the company had taken action to eliminate the practice of working on energized circuits, which was commonplace in the early days of industrial electrification. In the early 1990s, leaders in the electrical safety improvement initiative outlined a strategy for establishing a culture for long-term continuous improvement in electrical safety. The elements of that strategy, shown in Table 1.3, describe organizational culture intent on long-term impact on preventing electrical incidents and injuries. The culture and continuous improvement strategy described in these papers and nurtured for more than 20 years resulted in significant improvement in reducing the severity and frequency of electrical injuries in the company. Most dramatic is the impact on the frequency of fatalities from electrical energy. In the 1980s and early 1990s prior to 1993, fatalities from electrical energy were occurring on average every 33 months. The chart in the figure represents a global workforce of employees and contractors that ranged from 80,000 to 120,000 during this period. From 1993 through 2016, there were zero fatalities from electrical hazards in company facilities.

The exposure to hazardous electrical energy increased due to the increasing application of electrical technologies for energy, control, and communications. The significant change was the shift in the electrical safety culture driven by the continuous improvement environment that looked beyond simple compliance with regulations and standards.

A Task Example

In this example, the power system for an industrial manufacturing facility included five 12-kV feeders and 13 substations configured to enable 100%

redundancy to allow maintenance access without disrupting production. The isolation of one substation involved 36 lockout points, including eight 12-kV circuit breakers, eight 12-kV potential transformers, six 480-V circuit breakers, six 480-V switches, five 480-V tie breakers, two 125-V DC control circuits, and one 120-V AC station power circuit. Nine sets of temporary safety grounds were applied for personnel protection. After establishing an Electrically Safe Work Condition (ESWC) for this substation, the supervising engineer recognized that the shutter in the rear of a 12-kV circuit breaker cell presented a residual risk. The cell shutter was an engineering control providing a barrier covering the 12-kV stationary connection contacts when the 12-kV circuit breaker was removed from the cell. A worker could easily lift the shutter in an empty cell, creating a dangerous exposure to bare, unguarded 12-kV circuit parts. After achieving the original ESWC, the supervising engineer asked the following questions:

1) Where is the hazard?
2) What have I done to control exposure to the hazard?
3) How can the controls fail?
4) Will additional controls further reduce the risk of exposure?

These questions demonstrate the power of creating a lens of system safety to illuminate the hidden danger. A previously unrecognized significant risk of exposure to lethal hazardous energy was avoided by asking the above simple questions in collaboration with the maintenance crew. This residual risk was controlled by adding the shutter lever as an additional lockout point to the ESWC plan.

An organization's electrical safety program is typically an integration of regulations and standards that establish minimum requirements for the safe design, installation, operation, and maintenance of electrical systems. By creating a "lens of system safety," electrical safety leaders can be vigilant in identifying and further reducing residual risk.

The above example shows how the lens of system safety was used during the execution of a work task with potential exposure to hazardous energy. The lens can be used throughout the life cycle of a project. Hazard identification, assessment of risk, and selection of risk controls during the design phases of the life cycle are generally more effective and less costly than during the post-design phases, as design changes are permanent and less dependent on human performance during the operational phases of the life cycle. Other application opportunities include incident investigations, equipment specification and procurement, safety audits, and pre-job planning.

The rigorous methodologies of system safety have proven effective in managing the risks of technological mishaps with catastrophic consequences. Significant further reduction in occupational electrical injuries and fatalities will require a culture shift in safety management. Compliance-based programs must shift from compliance only to incorporate risk management and other concepts derived from system safety. Electrical safety leaders must commit to continual learning of the most effective approaches to safety management.

Adding risk management components to compliance involves an in-depth understanding of hazard identification, risk assessment, and risk reduction to supplement the results in injury prevention afforded by a compliance-based approach. A pure application of system safety to common risks, such as electrical injuries, is not practical or reasonable due to the resources and competencies needed. However, lessons from system safety can be used to identify and control risks that may remain after compliance with regulations and standards is achieved.

Significant improvement in preventing injuries and fatalities from exposure to electrical energy is possible and achievable. Electrical safety leaders can use the lens of system safety to highlight ways to address the residual risk left unmitigated by a compliance-only approach to electrical safety. Using a lens of system safety can change the paradigm of compliance and the opportunities to integrate the lessons of system safety into existing electrical safety regulations, standards, and programs to prevent electrical injuries and fatalities further.

Crosstalk: Questions for Thought and Discussion

1. How do you and your organization view regulations, codes, and standards? Are they minimum or optional requirements in your electrical safety program?

2. Have you ever heard someone say, "If OSHA doesn't require it, why should we do it?" What does this say about the electrical safety culture in your organization?

3. How does your organization describe acceptable risk?

4. How do you know you have achieved acceptable risk?

5. How do you know if what is considered acceptable risk could be reduced with reasonable effort?

Further Reading

Anderson, M. and Denkl, M. The Heinrich triangle—Too simplistic a model for HSE management in the 21st century? *Proc. Int. Conf. Health Safety Environ.* (December 2010), Rio de Janeiro, Brazil, Society of Petroleum Engineers.

ANSI/ASSE Z590.3. Prevention through design guidelines for addressing occupational hazards and risks in design and redesign processes, Des Plaines, IL, USA, American Society of Safety Professionals.

ANSI/ASSP Z10. Occupational health, and safety management systems. Des Plaines, IL, USA, American Society of Safety Professionals.

ANSI/ASSP Z590.3. Guidelines for applying prevention through design in design and redesign processes. Des Plaines, IL, USA, American Society of Safety Professionals.

Barr, N.J. (2015). *System Safety Engineering and Risk Assessment*. Boca Raton, FL: CRC Press.

Benton, J. ASSE & OSHA concentrate collective efforts on risk-based approach. *EHS Safety News* (26 October 2012), retrieved from https://ehssafetynewsamerica.com/2012/10/26/asse-osha-concentrate-collective-efforts-on-risk-based-approach-in-oil-gas-industry/.

Cawley, J.C. and Brenner, B.C. (2013). Occupational electrical injuries in the U.S., 2003–2009. *IEEE Industry Applications Magazine* 19 (3): 16–20.

Dezfuli, H., Benjamin, A., Everett, C. et al. *NASA System Safety Handbook, – Vol 2: System Safety Concepts, Guidelines, and Application Examples*. US National Aeronautics and Space Administration NASA/SP-2014-612, version 1.0 (1 November 2014).

Ericson, C.A. (2005). *Hazard Analysis and Techniques for System Safety*. Hoboken, NJ: Wiley.

Floyd, H.L. (2024). Through the lens of systems safety: the limitations of a compliance-based safety culture and opportunities to reduce electrical injuries. *IEEE Industry Applications Magazine* 30 (3): 12–17.

Floyd, H.L. and Cole, B.C. (2015). Twenty years later: creating a continuous improvement environment for electrical safety. *IEEE Transactions on Industry Applications* 51 (3): 2633–2640.

Floyd, H.L., Cole, B.C., Doughty, R.L. et al. (1994). Creating a continuous improvement environment for electrical safety. *IEEE Transactions on Industry Applications* 30 (3): 543–552.

ISO 45001. Occupational health and safety management systems, international standards organization, Geneva, Switzerland.

Kolak, J.J. Using system safety techniques to perform hazard analysis. *Conference Record of American Society of Safety Engineers Safety 2012 Conference* (3–5 June 2012), Denver, CO.

Krause, T., Bell, K., and Libby, D. Serious injury and fatality prevention – 10 years later (25 May 2020). https://krausebellgroup.com/serious-injury-and-fatality-prevention/.

Lewis, C.L., Haug, H.A. (2006). *The System Safety Handbook*. Self-published, retrieved from the internet at https://www.aerohabitat.eu/uploads/media/14-04-2006_-_Lewis_-_Haug__System_safety_handbook__0.3MB_.pdf.

Liberty Mutual Research Institute for Safety. *Work Related Electrical Injuries: Study Sparks New Insights*, vol. 13, no. 3, Liberty Mutual Research Institute for Safety, Hopkinton, MA: 2010.

Lyon, B. and Popov, G. (2018). *Risk Management Tools for Safety Professionals.* Parkridge, IL: Am. Society of Safety Professionals.

Manuele, F. (2013). Preventing Serious Injuries & Fatalities: Time for a Sociotechnical Model for an Operational Risk Management System. *Journal for Professional Safety* (May 2013).

Manuele, F. (2020). *Advanced Safety Management, Focusing on ANSI Z10, ISO 45001 and Fatality Prevention*, 3e. Hoboken, NJ, USA: Wiley.

Mendeloff, J. and Staatsky, L. (2014). Occupational fatality risks in the U.S. and the U.K. *American Journal of Industrial Medicine* 57: 4–14.

Peterson, D. Why safety is a people problem. *EHS Today* (31 March 1996), retrieved from https://www.ehstoday.com/archive/article/21907885/dan-petersen-why-safety-is-a-people-problem.

Peterson, D. and Williamsen, M. (2007). The culture of safety. *Journal for Professional Safety* March 2007: 17–19 and 57–59.

Stevans, R. (2004). *System Safety for the 21st Century*. Wiley.

U.S. Bureau of Labor Statistics. Injury, illness and fatalities. https://www.bls.gov/iif/latest-numbers.htm.

Walter, L. Facing the unthinkable – fatality prevention in the workplace. *EHS Today* (31 December 2007). https://www.ehstoday.com/safety/article/21904194/facing-the-unthinkable-fatality-prevention-in-the-workplace.

Wyzga, R.E. and Lindroos, W. (1999). Health implications of global electrification. *Annals. New York Academy of Sciences* 888: 1–7.

2

What We Can Learn from a Nuclear Mishap

Before Regulations and Standards

In the past three decades of the 20th century and the first two decades of the 21st century, occupational safety in most organizations in North America centered on conformance with government regulations and industry consensus standards first published in the 1990s. Before 1990, one would not ask the questions, "What does OSHA require?" or "What guidance do industry consensus standards provide?" These resources were yet to be published. Before 1990, the technical committees of the IAS Industrial and Commercial Power Systems (I&CPS) Department had evolved as centers of excellence for standards and guidance for industrial power system engineering. In 1984, the I&CPS Maintenance, Operations, and Safety (MOS) Subcommittee had three initiatives underway: advocacy for an OSHA regulation on electrical safety requirements for general industry, expansion of the recently published NFPA 70E *Electrical Safety Requirements for Employee Workplaces* to incorporate electrical safe work practices, and creation of a new IEEE standard that would provide recommendations on maintenance, operation, and safety of industrial and commercial power systems, IEEE 902, *Guide for Electrical Safety in Industrial and Commercial Power Systems.* In the absence of regulations and standards, subcommittee members provided leadership in hazard identification, risk assessment, and application of engineering designs and safe work practices to reduce the risk of injuries from hazardous electrical energy. In 1969, Ralph Lee published one of the first IEEE papers on industrial electrical safety. John Caddick and Thad Brown reviewed Bill Jordan's work in developing novel methods to protect workers from arc flash burns. Bernard Whittington and Bruce McClung demonstrated advancements in applying engineering designs to enhance power system reliability and worker safety. McClung and John Gallagher published a collection of industry best practices to establish an industrial operations electrical safety program. In 1989, the MOS Subcommittee held a ½ day symposium on the prevention of arc flash

Hidden Risk in Occupational Electrical Safety: Through the Lens of System Safety,
First Edition. H. Landis "Lanny" Floyd.

injuries at the IAS Annual Meeting. Ralph Lee had been involved in the early planning but passed away in 1987. Symposium speakers included Thad Brown, Lanny Floyd, Bill Jordan, Thad Masters, Bruce McClung, and John Moore. The symposium was the first industry forum to bring together thought leaders on the prevention of arc flash injuries in industrial power systems. For several decades before the first electrical safety regulations and industry standards, the ground-breaking collaborative work of the pioneering electrical safety leaders in the MOS Subcommittee laid the foundation for electrical safety federal regulations, industry standards, and technology advancements that were to come.

Learning from System Safety

On March 28, 1979, an accident began unfolding at the Three Mile Island nuclear-generating plant near Harrisburg, PA. The accident had the potential to evolve into one of the most catastrophic manmade events in history. Fortunately, the accident was controlled before the release of radioactive materials. U.S. President Jimmy Carter chartered a commission to investigate the accident. The executive summary of the commission's report included these statements: "We note a preoccupation with regulations.... we are convinced that regulations alone cannot assure safety... This commission believes that it is an absorbing concern with safety that will bring about safety – not just the meeting of narrowly prescribed and complex regulations."

> System safety: The application of engineering and management principles, criteria, and techniques to achieve acceptable mishap risk, within the constraints of operational effectiveness and suitability, time, and costs, throughout all phases of the system life cycle.
>
> –International System Safety Society

System safety has its roots in the 1940s, with the research and development of nuclear weapons and rapidly expanding military and commercial aviation industries. Designers, manufacturers, and pilots were pushing the envelope in technology. As airplanes became more sophisticated, the cost of mishaps escalated. The aviation industry recognized that the practice of analyzing mishaps after the fact was becoming unacceptable in terms of human safety and financial loss. In his paper, "Using Systems Safety Techniques to Perform Hazard Analysis," Kolak noted, "The costs associated with damaging expensive fighter jets and the development of nuclear energy (where a single system failure was unacceptable) contributed to the concept that hazards must be anticipated and controlled

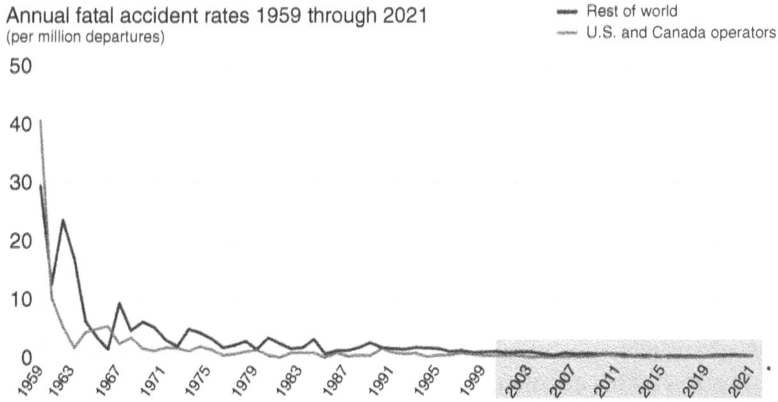

(a)

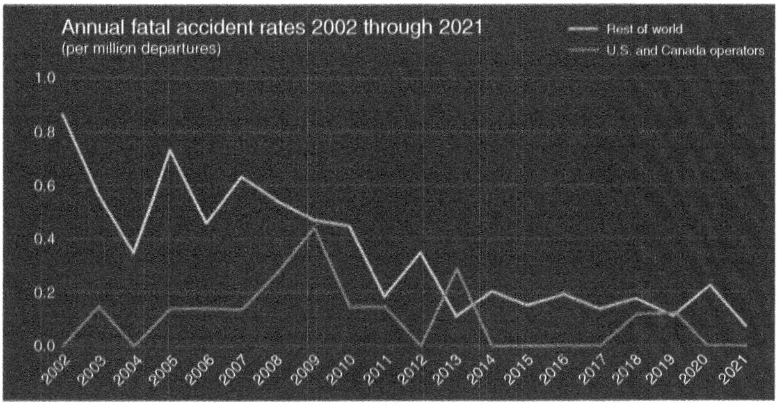

(b)

Figure 2.1 (a) The trend in fatal accidents in commercial jet aviation, 1958–2021. (b) The vertical axis in (a) is expanded to show the trend in fatal accidents in commercial jet aviation, 2002–2021.

before even a single loss occurs." The extraordinary improvement in commercial aviation safety shown in Figure 2.1a, b is primarily attributed to the development and application of systems safety.

The extraordinary results of applying system safety methods to control the extraordinary potential catastrophic consequences of 20th-century technology advancement can be applied to create "an absorbing concern with safety" to manage the risk to workers exposed to hazardous electrical energy. In other words, we can create and look through the lens of system safety to see opportunities to reduce the risk of electrical incidents and injuries.

A Culture of Compliance

As shown in Figure 2.2a, over the past 40 years, there has been a significant improvement in reducing occupational fatalities from exposure to electrical hazards. However, Figure 2.2b that shows the downward decline has been flat for the past ten years. The flattening of the downward trend is not unique to fatalities from exposure to hazardous electrical energy. Figure 2.3 shows that the trend in all occupational fatalities in the United States has also flattened.

These two charts indicate that the underlying causes of the flattened trend in electrical fatalities are likely not unique to electrical safety. Compliance is about adherence to regulations and rules. For electrical safety in the United States, compliance involves asking questions such as "What does OSHA require?" and "What do industry consensus standards require?" Both OSHA regulations and industry consensus standards provide minimum requirements for worker protection. Compliance with minimum requirements may result in residual or unidentified risk that could be further reduced.

Exploring electrical safety through the lens of systems safety involves understanding how hazard identification, risk assessment, and risk reduction can supplement the results in injury prevention afforded by a compliance-based approach.

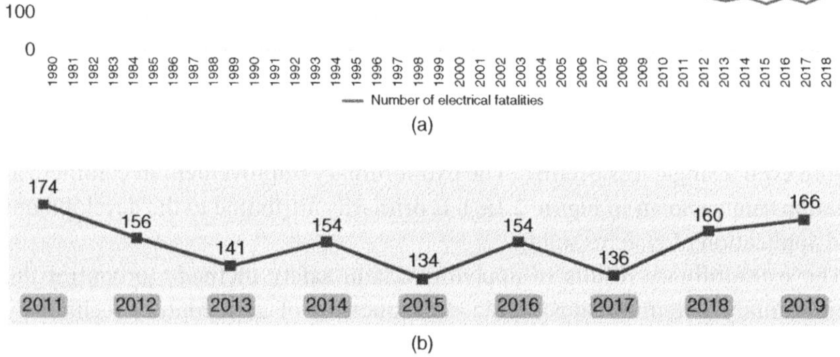

Figure 2.2 (a) The trend in occupational fatalities from exposure to electrical hazards in the United States, 1980–2018. (b) The trend in occupational fatalities from exposure to electrical hazards in the United States, 2011–2019. From Electrical Safety Foundation International.

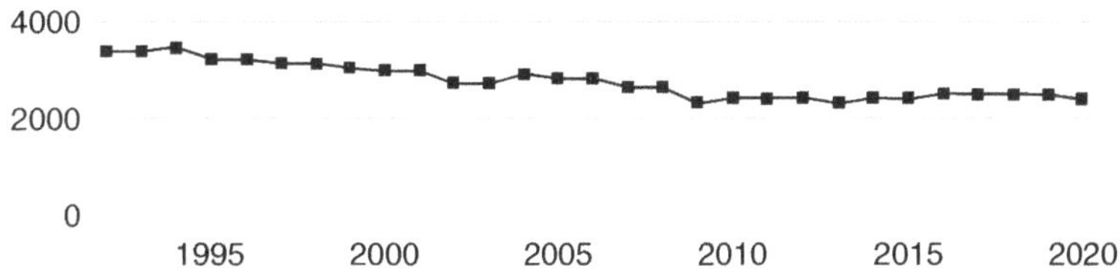

Figure 2.3 The trend in occupational fatalities from all causes in the United States.

After achieving compliance with safe work practices, applying risk management techniques includes exploring answers to questions such as "How can the risk controls fail? What are the consequences if one or more controls fail? Are the potential results acceptable?" These questions are fundamental to systems safety. In the context of possible exposure to hazardous electrical energy, asking and answering these questions can help see beyond compliance with minimum requirements to how hidden hazards can be identified and residual risk can be further reduced.

Applying Systems Safety Concepts

Figure 2.4 shows a collage of electrical safety regulations and standards at the center. The topics surrounding the regulations and standards are either absent or not well developed in the regulations and standards. Exploring these topics through a lens of system safety can uncover opportunities for advancements in hazard identification, risk assessment, risk control evaluation, the hierarchy of risk controls, prevention through design, human and organizational performance, and safety management systems – all topics embedded in systems safety.

An Evolution in Safety Management

Outside the United States

In 1994, the United Kingdom demanded the need for construction companies, project owners, and architects to address safety and health during the design phase of projects. Companies there responded with positive changes in management practices to comply with the regulations. Australia developed the Australian National Occupational Health and Safety (OHS) Strategy 2002–2012, which set eliminating hazards at the design stage as one of the five national priorities. In 2008, the EU launched a two-year campaign, with a focus on expanding the application of risk management in occupational safety. In announcing the campaign, it stated that, "*Risk assessment is the cornerstone of the European*

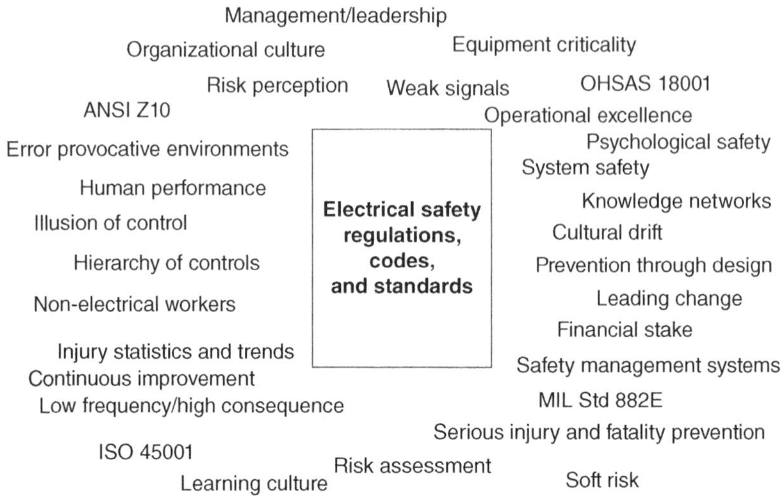

Figure 2.4 Inside the box are electrical safety-related regulations, codes, and standards. Outside the box are examples of safety management topics essential to electrical safety, but not included or not significantly discussed in electrical safety regulations, codes, and standards.

approach to prevent occupational accidents and ill health. If the risk assessment process – the start of the health and safety management approach – is not done well or not at all, the appropriate preventative measure are unlikely to be identified or put in place."

The UK Leaps Forward

Mendleoff and Staatsky studied occupational fatalities in the United States and United Kingdom and showed that the overall occupational fatality rate in the United Kingdom was approximately one-third that of the United States. As shown in Table 2.1, from their study, the fatality rate from exposure to electrical energy was about one-fourth that of the United States. One factor contributing to this difference is the safety management culture in the United Kingdom that places more emphasis and resources on risk assessment and application of a hierarchy of controls than the compliance approach common in the United States.

The NIOSH Prevention Through Design National Initiative

In 2007, the U.S. National Institute for Occupational Safety and Health (NIOSH) announced the National Initiative on Prevention through Design. The initiative is based on applying risk management concepts to identify opportunities to reduce risk beyond compliance with existing regulations and standards.

Table 2.1 A comparison of occupational fatality rates between the United States and the United Kingdom (from Mendleoff and Staatsky).

	UK	USA	
Event type	**Rate**	**Rate**	**Ratio**
Contact with moving machinery	0.06	0.2	3.3
Struck by moving, including flying/falling object	0.13	0.38	2.9
Struck by moving vehicle	0.11	0.43	3.9
Strike against something fixed or stationary	0.025	0.006	0.2
Injured while handling, lifting, or carrying			
Slips, trips, or falls on the same level	0.018	0.055	3.1
Falls from a height			
Up to 2 m			
Over 2 m	0.17	0.50	2.9
Height not stated			
Trapped by something collapsing/overturning	0.06	0.07	1.2
Drowning or asphyxiation	0.028	0.055	2.0
Exposure to, or contact with, a harmful substance	0.012	0.13	10.8
Exposure to fire	0.011	0.053	4.8
Exposure to an explosion	0.013	0.051	3.9
Contact with electricity or electrical discharge	0.037	0.14	3.8
Injured by an animal	0.015	0.025	1.7

The DuPont Company

During the mid-1980s, electrical safety leaders in the DuPont Company began collecting electrical incident and injury data from across the company's global operations. Electrical injuries were rare, so they were considered random events. The electrical safety leaders began to recognize that there may be a result of a systemic gap in the company's safety management system and brought it to top management's attention. In 1989, the management made a highly visible commitment to reducing the risk of injuries from electrical hazards to employees and contractors. Goals for sustainable improvement were established, financial support was provided, and dedicated people were empowered to reduce the likelihood of electrical incidents. We began crafting a focused application of the company's safety management expertise in highly hazardous processes to the unique hazards of electrical energy.

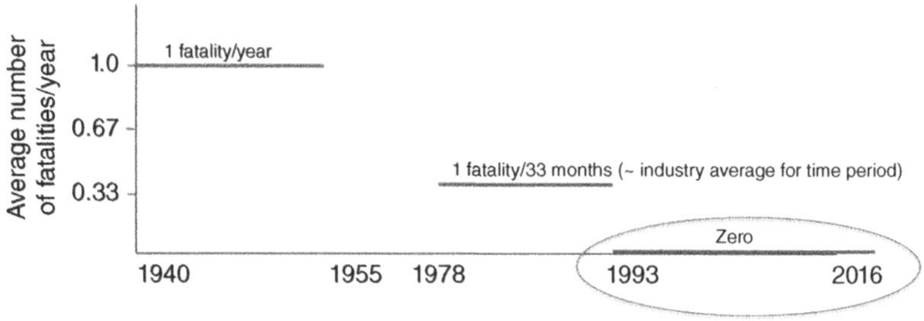

Figure 2.5 Improvement in occupational fatalities from exposure to electrical energy in the company's global operations. Employee and contractor workforce during the period was ~100,000.

In 1992, electrical safety leaders published the paper "Creating a Continuous Improvement Environment for Electrical Safety," which outlined a strategy for integrating advanced concepts in risk management into the company's existing electrical safety program. The culture and continuous improvement strategy nurtured for more than 25 years resulted in significant progress in reducing the severity and frequency of electrical injuries in the company. Most dramatic is the impact on the frequency of fatalities from electrical energy. As shown in Figure 2.5, before 1993, fatalities from electrical energy were occurring on average every 33 months. The chart represents a global workforce of employees and contractors that ranged from 80,000 to 120,000. The fatality frequency was the industry average for the period. From 1993 to 2016, there were zero fatalities from exposure to electrical energy in company facilities, demonstrating that significant and sustainable improvement is achievable.

Electrical hazards were not eliminated during this period, but instead increased due to the increasing application of electrical technologies in energy, control, and communication systems. What did change was the emphasis on management engagement, employee participation, near-miss incident learning, hazard identification, prevention through design, and enhancements to existing administrative controls and personal protective equipment (PPE).

Signaling a Shift from Compliance Only to Include Risk-Based Safety Management

In 2010, the assistant secretary of the U.S. Department of Labor, Dr. David Michaels, issued a statement to the OSHA staff: "Ensuring that American workplaces are safe will require a paradigm shift, with employers going beyond simply attempting to meet OSHA standards, to implementing risk-based workplace injury and illness prevention programs." Although OSHA has not proposed any

Table 2.2 Risk-based questions that look beyond compliance.

After compliance with regulations and standards
• Where is the hazard?
• What controls are in place?
• How can the controls fail?
• Will additional controls reduce risk?

regulatory requirements to facilitate this shift, resources are available to help make this shift. Table 2.2 illustrates a way to think and act beyond simple compliance.

Conclusion and Recommendations

In 2012, the American Society of Safety Professionals applauded the efforts of OSHA, the Environmental Protection Agency, the Pipeline and Hazardous Materials Safety Administration, the Bureau of Safety and Environmental Enforcement, and the U.S. Coast Guard for their advocacy of risk-based enhancements to regulatory requirements.

The potential for exposure to hazardous electrical energy has increased due to the growth in applying electrical technologies for energy, control, and communications in industrial applications. The evolution in occupational electrical safety regulations and standards in the United States has contributed to a significant reduction in occupational electrical injuries. However, the approach taken in the United States has created a widely held paradigm that compliance will be sufficient. Other countries have demonstrated that compliance combined with advanced safety concepts to more effectively identify hazards, assess risks, and evaluate risk control effectiveness can impact further reductions in injuries and fatalities from exposure to electrical hazards. Early electrical safety leaders exhibited exceptional competency in hazard identification and application of risk management techniques. They were likely influenced by the events and concerns that led to the development of system safety.

Significant further reduction in occupational electrical injuries and fatalities will require a culture shift in safety management. Compliance-based programs will need to shift from compliance only to incorporate risk management and other concepts derived from systems safety. Electrical safety leaders will need to commit to continual learning of the most effective approaches to safety management. Adding risk management components to compliance involves an in-depth understanding of hazard identification, risk assessment, and risk reduction to supplement the results in injury prevention afforded by a compliance-based approach. Questions like those shown in Table 2.2 need to be asked after achieving compliance with the requirements in regulations and standards.

Crosstalk: Questions for Discussion

1. If you received notice that someone in your organization had suffered a fatal electric shock this morning, what would you do differently tomorrow?
2. Why have you not already done what you answered to the above if your organization has not experienced a question?
3. Is it better to do whatever is necessary to make sure it does not happen again, or to make sure it did not happen the first time?
4. If your organization has not experienced a serious electrical injury recently, or perhaps never, is it due to the quality of your electrical safety program or is it because serious electrical injuries are rare?
5. Are you sure?

Further Reading

OSHA 1910, Subpart S, *Electrical*, retrieved from the internet at https://www.osha.gov/laws-regs/regulations/standardnumber/1910.

NFPA Standard 70E (1988). *Standard for Electrical Safety in the Workplace.* Batterymarch, MA: National Fire Protection Association.

IEEE Standard 902 (1998). *IEEE Guide for Maintenance, Operation, and Safety for Industrial and Commercial Power Systems.* Piscataway, NY: Institute of Electrical and Electronics Engineers.

Floyd, H.L., Halpin, M.S., and Saunders, L.F. An overview of the IEEE color books, *Conference Record of the 2000 IAS Annual Meeting*, Rome, Italy.

Lee, R., Industrial electrical safety, *IEEE Spectrum*, June 1971.

Caddick, J. and Brown, T. (1980). Operations and Safety for Electrical Power Systems. *Chemical Engineering Magazine, April* 21: 145–149.

Whittington, B. and McClung, L.B. (1974). Ground Fault Tests on High-Resistance Grounded 13.8-kV Y Electrical Distribution System of Modern Large Chemical Plant - Part 1. *IEEE Transactions on Industry Applications* 1a-10 (5): 601–617.

Whittington, B. and McClung, L.B. (1974). Ground fault tests on high-resistance grounded 13.8-kv y electrical distribution system of modern large chemical plant – part 2. *IEEE Transactions on Industry Applications* 1a-10 (5): 618–632.

McClung, L.B., Gallagher, J., Electrical Hotwork Safety Program, self-published, 1987, retrieved from the internet at http://site.ieee.org/ias-esafc/files/2016/03/McClung-and-Gallagher-Electrical-Hot-Work-Safety-Program.pdf.

Report of the President's Commission on the Accident at Three Mile Island, retrieved from the internet at http://large.stanford.edu/courses/2012/ph241/tran1/docs/188.pdf.

Lewis, C.L., Haug, H.A., The Safety Systems Handbook, self-published, retrieved from the internet at www.aerohabitat.org/link/2006/14-04-2006 - Lewis Haug, System safety handbook (0.3MB).

Kolak, J.J., Using system safety techniques to perform hazard analysis, *American Society of Safety Engineers Safety 2012 Conference*, Denver, CO June 3–5, 2012.

Statistical Summary of Commercial Jet Airplane Accidents Worldwide Operations 1959–2021, retrieved from the internet at http://www.boeing.com/news/techissues/pdf/statsum.pdf.

Floyd, H.L. (2015). A systems safety approach to occupational electrical safety. *IEEE Transactions on Industry Applications* 51 (2): 1284–1288.

National Safety Council Injury Facts, Trends in occupational fatalities, retrieved from the internet at https://injuryfacts.nsc.org/work/work-overview/work-related-fatality-trends/.

Brusso, B.C., Doan, D.R., and Floyd, H.L. (2022). The IAS electrical safety workshop: Celebrating 30 years of changing and advancing the electrical safety culture. *IEEE Industry Applications Magazine, May/June* 28 (3): 8–15.

Australian National Occupational Health and Safety (OHS) Strategy 2002–2012, Work Safe Australia, retrieved from the internet at https://www.safeworkaustralia.gov.au/system/files/documents/1702/nationalohsstrategy_2002-2012.pdf.

Manuele, F., "Preventing serious injuries & fatalities: Time for a sociotechnical model for an operational risk management system," *Journal for Professional Safety*, May 2013.

Mendeloff, J. and Staatsky, L. (2014). Occupational fatality risks in the U.S. and the U.K. *American Journal of Industrial Medicine* 57: 4–14.

Cole, B.C., Doughty, R.L., Floyd, H.L. et al. (1994). Creating a continuous improvement environment for electrical safety. *IEEE Transactions on Industry Applications* 30 (3): 543–551.

Floyd, H.L. and Cole, B.C. (2015). Twenty years later: Creating a continuous improvement environment for electrical safety. *IEEE Transactions on Industry Applications* 51 (3).

OSHA 3885. *Recommendations for Health and Safety Programs*, October 2016.

Benton, J, ASSE & OSHA concentrate collective efforts on risk-based approach, *EHS Safety News*, Oct 26, 2012, retrieved from https://ehssafetynewsamerica.com/2012/10/26/asse-osha-concentrate-collective-efforts-on-risk-based-approach-in-oil-gas-industry/.

ANSI/ASSP Z10-2019 (2019). *Occupational Health and Safety Management Systems.* American Society of Safety Professionals.

ISO 45001-2018 (2018). *Occupational Health and Safety Management Systems.* Geneva, Switzerland: International Standards Organization.

Manuele, F. (2020). *Advanced Safety Management, Focusing on ANSI Z10, ISO 45001 and Fatality Prevention*, 3e. Hoboken, NJ: Wiley.

Ericson, C.A. (2005). *Hazard Analysis and Techniques for System Safety*. Hoboken, NJ: Wiley.

Lyon, B. and Popov, G. (2018). *Risk Management Tools for Safety Professionals.* Parkridge, IL: Am. Society of Safety Professionals.

Barr, N.J. (2015). *System Safety Engineering and Risk Assessment*. Boca Raton, FL: CRC Press.

Floyd, H.L. (2023). Reflections on a 50 year journey in electrical safety. *IEEE Industry Applications Magazine* 29 (4): 59–65.

3

The Limitations of a Compliance-Based Safety Culture

Introduction

Have you ever heard someone say, "If OSHA or 70E doesn't require it, why should we do it?" Statements like this are characteristic of a compliance-based safety culture, which embraces the belief that simply complying with safety regulations, codes, and standards will achieve the desired safety performance. This approach, although limited, has resulted in long-term trends in reducing fatal and nonfatal occupational injuries in the United States. Emergence of the first OSHA laws in 1970 stimulated the development of industry consensus standards documenting safe work practices, such as NFPA 70E. A compliance-based culture has been effective in improving safety performance in the United States. Nearly 15,000 occupational fatalities from all causes occurred in the United States in 1970. With a workforce that has doubled since 1970, there were 5,486 occupational fatalities in 2022.

Since President Nixon signed into law the OSHA of 1970, workplace electrical safety in the United States has primarily focused on compliance with the minimum safe work practices in OSHA regulations and, after its first publication in 1979, compliance with requirements of the safe work practices in NFPA 70E, *Standard for Electrical Safety in the Workplace*. The requirements in NFPA 70E are also considered minimum electrical injury risk control measures.

Comparing the United States to Other Countries

For more than 40 years, the United States has seen significant reduction in both fatal and nonfatal LTIs from exposure to hazardous electrical energy; however, the downward trends have flattened in recent years. The flattening trend could be incorrectly interpreted that the United States may be at a theoretical limit in reducing serious electrical injuries. However, other countries have demonstrated

Hidden Risk in Occupational Electrical Safety: Through the Lens of System Safety,
First Edition. H. Landis "Lanny" Floyd.

significantly lower occupational fatality rates, including electrical fatality, than the United States. The report "Occupational fatality risks in the United States and the United Kingdom," by Mendleoff and Staatsky published in the *American Journal of Industrial Medicine* in 2014, showed that the overall occupational fatality rate in the United Kingdom was approximately one-third that of the United States. The fatality rate from exposure to electrical energy was about one-fourth that of the United States. The study did not attempt to determine what has contributed to the difference in safety performance; however, the safety management regulations in the United Kingdom place more emphasis and resources on risk assessment than the compliance-based approach common in the United States.

Low Frequency/High Consequence

Based on the Bureau of Labor Statistics data, less than 0.2%, or less than 1 in 500, LTIs in the United States in 2019 were attributed to exposure to hazardous electrical energy. The relative infrequency of electrical injuries can create an illusion impacting the estimation of the likelihood of exposure to hazardous energy when performing a task. Incidents that are low in frequency can also mean that a company or an organization does not have sufficient internal data on electrical injuries to be statistically meaningful. Due to the inherent low-frequency occurrence of electrical injuries, an individual or a company may not recognize the potential for a fatal injury. Consider a task in which the likelihood of injury when a worker performs a task is viewed as unlikely. The worker and management may consider the risk acceptable and proceed to perform the task without any additional controls to reduce risk further. Now consider the worker may perform a similar task every day over a 20-year career, which could mean the worker could perform tasks with unlikely injury potentially 4,000 times during the worker's career. Now the "unlikely" incident may be up to 4,000 times more likely to result in serious injury or death sometime during the worker's career. Or consider a large company with many manufacturing facilities in which the task could be performed one million times/year. Now the "unlikely" incident based on assessing one task is one million times more likely to occur to a worker somewhere in the company's facilities. Underestimating the potential frequency of a serious incident could have tragic consequences.

Addressing Residual Risk

Residual risk is the risk that remains after risk controls are taken into account. The primary goal in risk-based occupational safety management is to reduce risk to ALARP. In conducting risk assessments, if the frequency of exposure is underestimated, control effectiveness overestimated, or potential severity of consequences underrecognized, the resulting residual risk may fall short of ALARP.

When establishing an ESWC, there may be residual risks, which, if assessed, could be further addressed with additional controls. Consider an example of a lineup of 15 kV industrial switchgear on which maintenance tasks of cleaning, inspections, and testing are to be performed. Is it feasible that an error could have occurred in establishing an ESWC? Is it feasible that an isolation point could have been missed, perhaps a potential transformer that could backfeed the 15 kV bus from a 120-V source? Is it feasible that the physical boundaries of the ESWC could be misunderstood by a worker? Could a worker be confused by "look-alike" equipment that is outside the safe work zone? Could a worker remove a cover exposing hazardous energy? To address these and other remaining risks, additional controls may be needed. Overestimating the effectiveness of the compliance-based controls could create an illusion of safety for workers.

The hierarchy of risk controls has been in safety management standards and literature for more than 50 years. With the exception of a control measure that permanently removes the hazard ensuring that no worker can be exposed to the hazard under any foreseeable circumstances, or one that substitutes a less severe hazard, other control measures serve to reduce the likelihood of exposure but do not change the potential severity of exposure to a hazard. The control measures of administrative controls and PPE are vulnerable to human error that reduces the effectiveness of control measures. The relatively infrequent occurrence of electrical injuries as compared to other injuries from other hazards in the workplace can create a false sense of immunity from the severe consequences of contact with electrical energy.

Having personnel competent in risk assessment methods and an ongoing robust process to facilitate identification, quantification, and reduction of risk is essential to achieving risk ALARP. Safeguarding workers from the pitfalls discussed earlier requires investment in time and money to develop competency in risk assessment and mitigation tools and methods.

The Three Mile Island nuclear power station accident in 1979 highlighted the limitations of a compliance-based safety culture. The *Report of the President's Commission on the Three Mile Island Nuclear Accident* states that

> We note a preoccupation with regulations.... we are convinced that regulations alone cannot assure safety... This commission believes that it is an absorbing concern with safety that will bring about safety – not just the meeting of narrowly prescribed and complex regulations.

Looking for Hidden Risk

After compliance with electrical regulations, codes, and standards, ask these questions:
- Where is the hazard now?
- What controls are in place to limit exposure?
- How can the controls fail?
- What additional controls will reduce risk?

Figure 3.1 Questions that can help identify hidden danger.

If compliance with regulations is not sufficient in nuclear safety, could this learning also apply to electrical safety? How can learnings to control the extraordinary potential catastrophic consequences of a nuclear incident be applied to create "an absorbing concern with safety" to manage the risk to workers exposed to hazardous electrical energy?

Exploring the residual risks that may remain after compliance with regulations and standards involves having a heightened understanding of hazard identification, risk assessment, and risk reduction to supplement the results in injury prevention afforded by a compliance-only approach. Employers and workers and organizations must recognize that simple compliance with the minimum requirements of regulations and industry standards may have residual risk that may not be ALARP. After achieving compliance with safe work practices, applying risk management techniques includes exploring answers to the questions shown in Figure 3.1. When there is possible exposure to hazardous electrical energy, asking and answering these questions can help move beyond compliance with minimum requirements to see hidden danger and how residual risk can be reduced further. Electrical safety leaders can explore these questions throughout the life cycle of a facility or system, including initial design, equipment specification and procurement, task planning, job safety analysis, and incident investigations.

The National Safety Council (NSC), the ASSP, and OSHA have partnered to provide resources to help organizations complement a traditional compliance-based approach with one that more effectively utilizes risk-based tools to move beyond simple compliance.

We have the opportunity to explore the "hidden danger" that may remain after simple complying with regulations, codes, and standards. In order to achieve full potential in injury and fatality reduction, electrical safety practitioners will need to invest time and energy to develop competency in moving beyond compliance. A lens of system safety will help illuminate the hidden danger.

As illustrated in Figure 3.2, an organization's electrical safety program requires integration of regulations, codes, and standards that establish minimum requirements for the safe design, installation, operation, and maintenance of

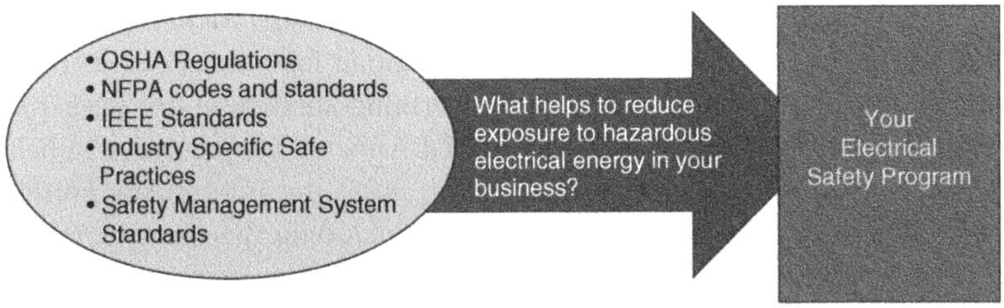

Figure 3.2 The electrical safety program requires integrating the requirements of multiple regulations, codes, and standards. Image created by the author.

electrical systems. By consciously looking beyond compliance, electrical safety leaders can be vigilant in identifying hidden danger and further reducing residual risk throughout the life cycle of a facility.

Applying Critical Thinking

> Critical thinking can be described as an ability to question; to acknowledge and test previously held assumptions; to recognize ambiguity; to examine, interpret, evaluate, reason, and reflect; to make informed judgments and decisions; and to clarify, articulate, and justify positions. When moving from a compliance-based culture to a risk-based culture, embracing critical thinking is essential to discovering opportunities hidden in residual risk. Accurate identification of hazards, assessment of risk, and evaluation of risk controls are essential to moving beyond simple compliance with electrical safety regulations and standards. If hazard identification and risk assessment are not done properly, or at all, the proper and most effective risk controls will often not be applied. This is especially true for electrical incidents as the very low frequency of electrical injuries, compared to all occupational injuries, can blind organizations and individuals to the potential for serious or fatal electrical injuries. Risk can be underestimated, and the effectiveness of controls overestimated. Risk determination is a function of incident consequences and the likelihood of ______. "An incident" may not be the right word to fill in the blank. A better choice may be "exosure to hazardous electrical energy."

There has been increasing emphasis that the application of risk assessment methodologies is needed to further reduce serious injuries and fatalities in the United States. In 2010, David Michaels, the assistant director of the US Department of Labor, stated that employers needed to move beyond simple compliance with regulations by incorporating risk-based approaches to complement compliance.

In 2016, OSHA published recommendations in OSHA 3885, *Recommendations for Safety and Health Programs*, that included a guide for risk assessments. In 2015, NFPA 70E added requirements for performing shock and arc flash risk assessments in task planning and also enhanced Annex F as a resource to help implement electrical hazard risk assessment. In 2025, the Intersociety Forum, a coalition of more than 35 safety organizations including the National Safety Council and the Americann Society of Safety Professionals, was formed to promote advanced safety management methods beyond simple compliance. The effectiveness of adding risk assessments to complement compliance with the minimum requirements provided in regulations and standards is dependent on the knowledge and competency of individuals performing the risk assessments. One variable impacting the quality of performing risk assessments includes the basic assumptions in defining risk. Brauer notes that there are many definitions of risk, especially looking beyond safety and health to other business concerns, such as project schedule, market share, and financial performance. However, it may be helpful to have common definitions for safety risk.

The addition of risk assessment requirements and resources in the 2015 revision of NFPA 70E included a definition for hazard and risk. Whether the risk of injury is from electric shock or arc flash, the source of harm, or hazard, is the same: hazardous electrical energy. However, there is a subtle, but perhaps significant difference in how NFPA 70E defines risk as compared to other safety management standards. As shown in Table 3.1 and Figure 3.3, the definition in NFPA 70E uses the likelihood of consequences from an exposure to a hazard, and ANSI/ASSP Z10, ISO 45001, and OSHA publication 3885 use the likelihood of exposure to a hazard. Again, the difference is subtle.

Table 3.1 Comparing definitions of risk.

Standard	Definition of Risk
NFPA 70E-2024, *Standard for Electrical Safety in the Workplace*	A combination of the likelihood of injury or damage to health and the severity of injury or damage to health that results from a hazard
ANSI/ASSP Z10-2019, *Occupational Health and Safety Management Systems*	An estimate of the combination of the likelihood of a hazardous event or exposure(s), and the severity of injury or illness that may be caused by the event or exposures
ISO 45001-2018, *Occupational Health and Safety Management Systems*	Combination of the likelihood of an occurrence of a work-related hazardous event(s) or exposure(s), and the severity of injury or ill health that can be caused by the event(s) or exposure(s)
OSHA 3885, *Recommendations for Safety and Health Programs*	A product of hazard and exposure

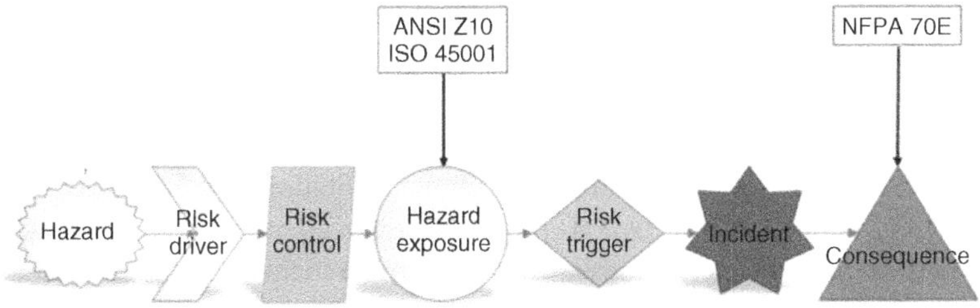

Figure 3.3 From hazard to consequence. Adapted from ANSI/ASSP Z10-2019.

How significant are these differences in these definitions? The model in Figure 3.3, adapted from ANSI/ASSP Z10-2019, provides a way to think about how a hazard can be transformed into an electrical injury. The stages of this model can be defined as follows:

- **Hazard:** Hazardous electrical energy
- **Risk driver:** Conditions that influence or increase the level of exposure
- **Risk control:** Actions or measures to reduce or mitigate risk
- **Hazard exposure:** In electrical environments, when a person is within the limited approach or arc flash boundaries
- **Risk trigger:** A condition or event that will cause an incident to occur
- **Incident:** An event with unwanted consequences
- **Consequence:** Unwanted injury or illness results

Electric shock, arc flash burns, and other secondary injuries, such as falls, are potential consequences of contact with or exposure to discharge of hazardous electrical energy. NFPA 70E uses the likelihood of injury, or consequence of an incident, in its definition of risk. ANSI Z10 and ISO 45001 use the likelihood of exposure to a hazard in its definition of risk. The assumptions made in performing risk assessments may impact decisions in risk control and mitigation.

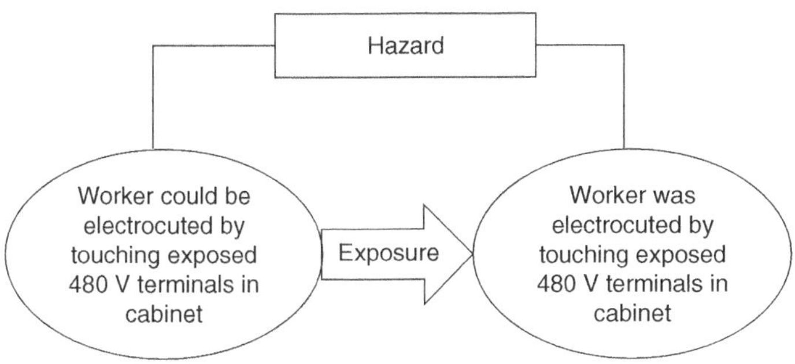

Figure 3.4 Same hazard – different states.

Figure 3.4 is adapted from Ericson and provides a different view of the relationship of hazard, exposure, and incident. In this illustration, the hazard in the state of potential injury is the same as the hazard in the reality of injury. The state is changed by exposure.

Assuming that the electrical equipment shown in Figure 3.5 is in an electrical safe working condition, the two workers do not need to be concerned with exposure to hazardous energy, measures to prevent an incident, or PPE to reduce the severity of injury if an incident were to occur. However, if the equipment is not in an electrical safe working condition and is energized at 480 V and the arc flash boundary is 34″, the workers are not in compliance with the requirements of OSHA 1910-330-335 and NFPA 70E. Both workers are within the limited approach boundary of 3′–6″, and the worker holding the test leads is within the restricted approach boundary of 1 foot. It is obvious that there is a likelihood of exposure to hazardous electrical energy. Both workers are within the arc flash and limited approach boundaries. But what is the likelihood of injury? If neither worker had ever experienced an electrical injury, their past experience

Figure 3.5 Workers performing diagnostic testing on energized 480-V equipment.

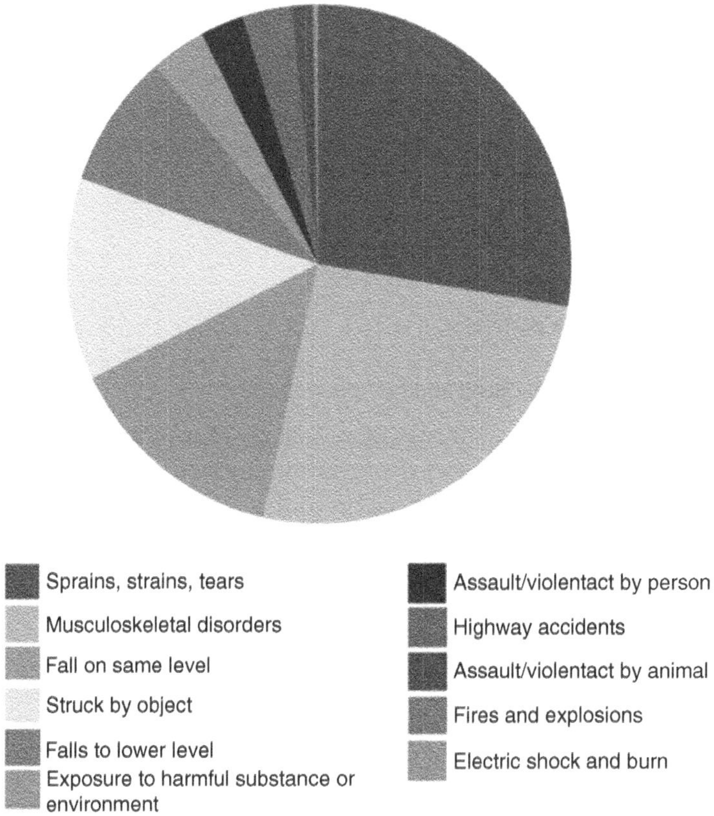

Figure 3.6 Select lost-time injuries in the United States for the year 2019. Based on US Bureau of Labor Statistics data.

could influence their decision that the likelihood of injury is low and the risk is acceptable.

Compounding the perception of risk of electrical injury is the fact that electrical injuries are relatively rare compared to all workplace injuries, as shown in Figure 3.6. LTI from contact with or exposure to electrical energy is less than 0.2% of all LTIs.

The exposure shown in Figure 3.5 could have been reduced by selecting design options that reduce the likelihood of exposure. Inherently safer design options could include the following:

- Isolating low-voltage (LV) equipment by placing 480-V devices in a separate enclosure
- Using IP20 touch safe terminals on all equipment

Although these options would not eliminate the hazard, they would reduce the likelihood of exposure to hazardous energy, leading to a reduced likelihood of an electric shock or arc flash incident resulting in serious injury.

Figure 3.5 shows the workplace as designed. Lower order controls applied in the workplace after design is completed in this example include the following:

- Worker training
- Task procedures
- Task planning
- Selection of test instrument
- Energized electrical work permit
- Voltage-rated gloves
- Arc-rated clothing, including head and face protection

The need to select all of these control measures would be obvious if the decision was based on the likelihood of exposure to hazardous electrical energy. If the workers determined that an electrical incident and electrical injury were unlikely, they may not have selected some or none of these control measures. Personnel performing electrical safety risk assessments should be aware of differences in assumptions made in defining risk and select the most appropriate approach to reduce risk to meet the organization's objectives.

An Unintentional Oversight – Nonelectrical Workers

For many organizations, the focus and improvement in electrical safety programs have largely addressed electrical workers whose primary work tasks involve construction, operation, and/or maintenance of electrical equipment. However, nearly 50% of workplace electrical injuries and fatalities are not electrical workers. Managers and administrators, painters, truck drivers, farm workers, grounds keepers, and gardeners are among the top 10 occupations having the most fatal electrical injuries. For these nonelectrical workers, the exposure to electrical hazards ranges from the use of common portable tools and appliances to unintentional contact with overhead power lines in the course of routine work activities.

We live in an electrical world, with nearly every aspect of modern business and commerce dependent on electrical technologies and interactions with tools, appliances, equipment, and systems that have some aspect of inherently hazardous electrical energy. Although the degree of risks ranges from little to great, it is the norm today that all people in the workplace have some risk of injury from electrical energy.

Exposure to electrical hazards remains a leading cause of occupational fatality. Cawley and Brenner reported it was the seventh highest cause of US workplace

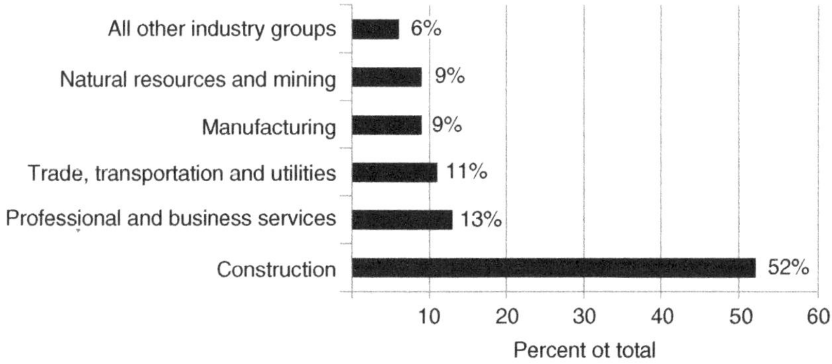

Figure 3.7 Workplace electrical deaths in the United States 2003–2009. Adapted from Cawley and Brenner [1].

fatality in a study spanning 2003–2009. Their study also reported that 18,460 workers were nonfatally injured from electrical hazards during the same period. As shown in Figures 3.7 and 3.8, the workplace environments suffering electrical fatalities and serious nonfatal injuries span the spectrum of industry, business, and commerce.

Since the passage of the OSHA in 1970, workplace injuries and fatalities, including those from electrical hazards, have trended downward. The emergence of more robust prescriptive industry standards, such as NFPA 70E, *Standard for Electrical Safety in the Workplace,* has been instrumental in accelerating the downward

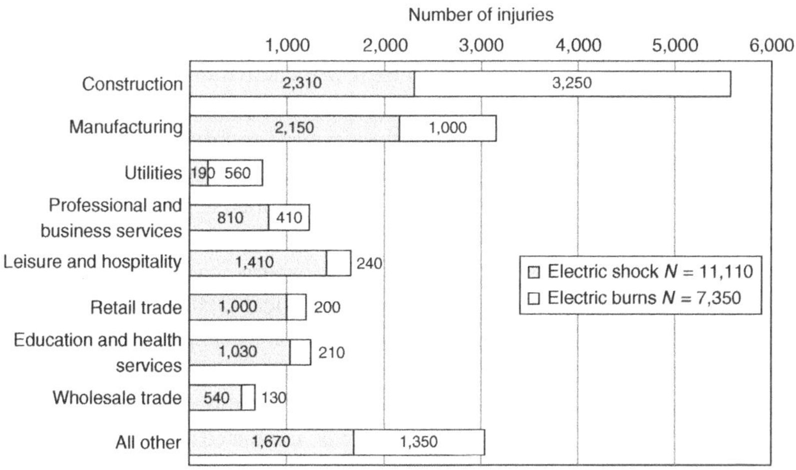

Figure 3.8 Workplace nonfatal electrical injuries in the United States 2003–2009. Adapted from Cawley and Brenner [1].

trend in injuries and fatalities from electrical hazards. However, misperception of who may be at risk for electrical injuries may be a limiting factor in achieving the potential for injury prevention and the application of best practices for preventing these injuries. Taylor et al. noted that managers and administrators, painters, truck drivers, farm workers, grounds keepers, and gardeners are among the top 10 occupations having the most fatal electrical injuries. It is essential that workplace electrical safety programs address electrical safety for all workers and not just workers whose job responsibilities involve working on or near energized electrical circuits. For these nonelectrical workers, the exposure to electrical hazards ranges from the use of common portable tools and appliances to unintentional contact with overhead power lines in the course of routine work activities.

This raises the questions: What is the workplace? What workers are at risk of electrical injury? Do existing electrical safety programs encompass all workplaces and all workers?

Article 90.1 of NFPA 70E states, *The purpose of this standard is to provide a practical safe working area for employees relative to the hazards arising from the use of electricity*. The scope description in Article 90.2 further narrows the application of the standard with this statement: *This standard addresses electrical safety requirements… for the practical safeguarding of employees during activities such as the installation, operation, maintenance, and demolition of electric conductors, electric equipment, signaling and communications conductors and equipment, and raceways*. Strict adherence to the boundaries in these statements would limit the application of this standard to workers whose job description specifically requires potential exposure to energized electrical equipment and systems. Could any of the requirements in the standard apply to other workers? What could be the impact if there were more effort to facilitate the application to other workers who have potential exposure to electrical hazards in their work activities?

DuPont is one of the oldest of the Fortune 500, incorporated in 1802. For the first 100 years, the company produced explosives used in farming, construction, hunting, and warfare. The manufacture and distribution of these products were inherently hazardous and fostered a robust and disciplined culture for managing safety for highly hazardous processes. At the end of the 19th century, electricity was introduced in the company's facilities, and the experience in managing safety was applied to this new technology. Of the 12 electrical fatalities that occurred between 1969 and 2016, 1 was a supervisor standing on a coal pile underneath an overhead power line, 3 were electricians in separate incidents working on equipment thought to be de-energized, 2 were carpenters who rolled a scaffold into an overhead electric line, 1 was a window washer who contacted an overhead power line while positioning a ladder against a building, 1 was a welder who attempted to troubleshoot a welding machine, 1 was a painter who came in contact with a high-voltage terminal on top of a substation transformer, 1 was a supervisor

Table 3.2 Job description of workers suffering fatal electrical injuries in DuPont, 1968–2016.

- Carpenter (2)
- Electrician (3)
- Electrical craft supervisor
- Electrical consultant
- Painter
- Powerhouse coal operations supervisor
- Sales and marketing representative
- Welder
- Window washer

providing troubleshooting assistance, 1 was an engineering consultant performing a visual inspection of electrical switchgear, and 1 was a sales representative who came in contact with an energized conductor in a customer's facility. Of the 12 fatalities, 3 were electricians, and only 5 could be characterized as workers whose job responsibilities would routinely place them near exposed electrical hazards (Table 3.2).

Although the degree of risks ranges from little to great, it is the norm today that all people in the workplace have some risk of injury from electrical energy. Equipment, appliance and tool designs, and installation requirements such as those found in NFPA 70, *National Electrical Code*, and IEEE/ANSI C2, *National Electrical Safety Code,* have continually evolved to reduce the risk of electrical injury from normal interaction by all workers. When the inherent safety features of electrical equipment or systems are not as intended due to damage, wear, improper installation, misapplication, misuse, or the hazard is unrecognized, the risk of electrical injury to the unsuspecting worker can significantly increase. Examples include a mobile television news crew that parks their vehicle underneath a power line and raises an antenna mast into the line, a carpenter who uses a power tool plugged into an extension cord with a damaged ground wire, a roofer handling a section of aluminum gutter near an overhead electric line, or a cover on an electric powered appliance is missing or unsecured creating a booby trap for an unsuspecting worker.

Research by Taylor et al., Cawley and Brenner, and Gammon and Vigstrol shows that while electricians, electrical apprentices, and electrical power installers are the highest at-risk groups, approximately half of electrical fatalities fall outside these electrical trades. As noted earlier, construction laborers, groundskeepers and gardeners, truck drivers, carpenters, managers and administrators, and painters are among the top 10 occupations sustaining the most fatal electrical injuries. Lombardi et al. have provided insight on work activities contributing to no fatal electrical injuries. Manual tasks characterized as plugging/unplugging, loading/unloading, cleaning/dusting, and tasks other than testing/checking power and turning on/off comprise 33.5% of activities contributing to nonfatal electrical injuries. The study by Taylor et al of fatal electrical injuries shows that vehicular and transportation operations, use of tools and machinery, and materials handling operations accounted for 31.3% of fatal electrical injuries during the period 1992–1999.

Overhead electric line contact is of particular concern, accounting for more than 40% of fatalities from electricity. Contact with overhead lines during activity other than electrical system installation and maintenance is the largest category of workplace electrocutions in the United States. Within this group, contact with a handheld object accounted for 22%, and contact indirectly through a piece of high-reaching mobile equipment, such as cranes, drilling rigs, mobile work platforms, antennas, irrigation rigs, and raised truck beds, accounted for 17%. Anything that can go up in the air has the potential to make unintentional contact with an overhead power line. This includes antenna masts on mobile equipment, water drilling rigs, crane booms, elevated work platforms, ladders, pipes, roof gutters, and building siding. Literally, any conductive object, material, or equipment that can be held or in contact with a person has the potential for extending the individual's reach to unintentionally contact an overhead electric line.

The use of ground fault circuit interrupters (GFCIs) has proven effective in significantly reducing fatalities associated with portable tools and appliances. Wells reported more than a 70% decrease in electrocutions involving portable tools and appliances in the 32-year period from 1970 to 2002. Requirements for GFCI installation were introduced in the 1968 National Electrical Code. Each edition since then has expanded the requirements for GFCI installation. The expanded application of GFCIs was likely a major contributor to the significant reduction in electrocution fatalities.

An organization intent on reducing the risk of electrical injury in its operations should consider actions in at least two areas: expand the use of GFCI protection and assess practices and procedures to prevent unintentional contact with overhead lines.

For individuals and organizations that may have perceived themselves as not being at risk of electrical injuries, electrical safety codes and standards may be a bit intimidating. For example, the owner of a small painting contractor firm that did not have an electrical expert on staff could conceivably have difficulty in interpreting and applying the requirements of NFPA 70E to their employees. With the major emphasis of most electrical safety codes and regulations being directed to subject matter experts or to workers routinely working on or near energized electrical conductors, tailoring electrical craft-specific requirements to apply to nonelectrical work environments may take some effort. Companies with established electrical safety programs are best positioned to critique and expand their program to more effectively include workers in nonelectrical crafts. Examples of questions in such a self-critique could include the following:

- Do we have a documented electrical safety program?
- Is it best characterized as an electrical safety program for workers in electrical crafts, or does it address the risk of electrical injury for all workers?
- Do workers use mobile equipment or handle material that could come in contact with overhead electric lines? How does our program control the movement of mobile equipment and material handling near overhead electric lines?
- Do workers use portable tools, appliances, and extension cords? How do we know GFCIs are being used?
- Do workers use portable tools and appliances in potentially damp or wet environments? How do we know GFCIs are being used?
- Do we have portable tools or appliances operating above 120 V that do not have GFCI protection? Are we applying the latest technology in GFCI protection?
- How disciplined is the organization in maintaining the integrity of enclosures and covers on electrical equipment, tools, and appliances?
- Are nonelectrical personnel included in arc flash protection programs? Do we understand who is at risk?

These are a few examples of questions, seen through a "lens of system safety," that could help identify opportunities to reduce the risk of electrical injuries to all workers. The questions could differ based on the industry, work environment, or maturity of the organization's electrical safety program.

Crosstalk: Questions for Discussion

1. Were you surprised to see that other countries have demonstrated significantly better performance in preventing occupational fatalities from exposure to hazardous electrical energy?

2. Do you and your organization believe that compliance with regulations and standards will result in acceptable risk? What makes you think that?

3. If you used a "lens of system safety" to audit a common task involving exposure to hazardous electrical energy, what could you learn?

4. Does compliance with the installation requirements of the National Electrical Code achieve an acceptable risk from electrical hazards?

5. How do you evaluate electrical injury risk in comparing different options when reviewing the design of an electrical installation?

Reference

1 Cawley, J.C. and Brenner, B.C. (2013). Analyzing on-the-job electrical injuries: a survey of selected U.S. occupational electrical injuries from 2003 to 2009. *IEEE Industry Applications Magazine* 19 (3): 16–20.

Further Reading

ANSI/ASSP Z10 (2019). *Occupational Health, and Safety Management System*. Des Plaines, IL: American Society of Safety Professionals.

Antman, S., Liggett, D., and Nolter, R. New Rules for Ground Fault Circuit Interrupters, *2011 IEEE IAS Electrical Safety Workshop* (30 January–4 February 2011), Toronto, ON.

Benton, J. "ASSE & OSHA concentrate collective efforts on risk-based approach," *EHS Safety News*, October 26, 2012, retrieved from https://ehssafetynewsamerica.com/2012/10/26/asse-osha-concentrate-collective-efforts-on-risk-based-approach-in-oil-gas-industry/.

Brauer, R.L. (2006). *Safety and Health for Engineers*, p645. John Wiley and Sons.

Brenner, B. Reinvigorating electrical safety. *2011 IEEE IAS Electrical Safety Workshop* (30 January–4 February 2011), Toronto, ON.

Ericson, C.A. (2005). *Hazard Analysis Techniques for System Safety*. Wiley.

Floyd, H.L. Through the lens of systems safety: the limitations of a compliance-based safety culture and opportunities to reduce electrical injuries, *IEEE Industry Applications Magazine*, May/June 2024.

Floyd, H.L. (2015). Facilitating application of electrical safety best practices to "other" employees. *IEEE Transactions on Industry Applications* 51 (2): 1348–1352.

Floyd, H.L. (2025). Aligning likelihood for clarity in electrical safety. *IEEE Industry Applications Magazine* 31 (3): 23–26.

ISO 45001 (2018). *Occupational Health and Safety Management Systems*. Geneva: International Standards Organization.

Lombardi, D.A., Matz, S., Brennan, M.J. et al. (2010). Etiology of work-related electrical injuries: a narrative analysis of workers' compensation claims. *Journal of Occupational and Environmental Hygiene* 6 (10): 612–623.

Manuele, F. (2013). Preventing serious injuries & fatalities: time for a sociotechnical model for an operational risk management system. *Journal for Professional Safety* 58 (5): 51–59.

Mendeloff, J. and Staatsky, L. (2014). Occupational fatality risks in the U.S. and the U.K. *American Journal of Industrial Medicine* 57: 4–14.

NFPA 70E. Standard for electrical safety in the workplace.

OSHA 3885 (2016). Recommendations for safety and health programs.

Peterson, D. Why safety is a people problem, *EHS Today*, March 31, 1996, retrieved from https://www.ehstoday.com/archive/article/21907885/dan-petersen-why-safety-is-a-people-problem.

Report of the President's Commission on the Three Mile Island Nuclear Accident.

Shutterstock image 3403403, used with permission.

Taylor, A.J., McGwin, G. Jr., Valent, F., and Rue, L.W. III, (2002). Fatal electrocutions in the United States. *Injury Prevention* 8 (4): 306–312.

Wells, J. GFCIs – thirty years and counting. *2003 IEEE IAS Electrical Safety Workshop* (4–7 February 2003), Houston, TX.

4

A Case History of Breakthrough Performance

Introduction

In 1989, the management of DuPont, a global science and technology company, made a highly visible commitment to reduce the likelihood and severity of injuries to employees and contractors from electrical hazards. Goals for sustainable improvement were established, financial support provided, and dedicated people empowered to reduce the likelihood of electrical incidents, injuries, and fatalities, with the intent to accomplish a step change in electrical safety performance, as was done in the mid-1950s. At that time, the company had taken action to eliminate the practice of working on energized circuits, which was commonplace in the early days of industrial electrification.

A Strategy for Change

In 1990 and 1992, several leaders in the company's electrical safety improvement initiative collaborated on two award-winning papers presented at the annual IEEE IAS Petroleum and Chemical Industry Conference and subsequently published in *IEEE Transactions on Industry Applications*. The first paper, *Maintaining Safe Electrical Work Practices in a Competitive Environment*, was presented at the 1990 IEEE IAS Petroleum and Chemical Industry Conference in Houston, Texas. This paper described the company's concern for improving electrical safety performance and the creation of an organizational infrastructure to enable and support changes to better manage electrical hazards in company facilities and operations.

The second paper, *Creating a Continuous Improvement Environment for Electrical Safety*, was presented at the 1992 conference in San Antonio, Texas. This paper outlined a strategy for establishing a culture for long-term continuous improvement in electrical safety. The elements of that strategy, shown in Table 4.1,

Hidden Risk in Occupational Electrical Safety: Through the Lens of System Safety,
First Edition. H. Landis "Lanny" Floyd.

Table 4.1 Elements of the strategy described in the paper.

• Understand the business consequences of electrical incidents
• Engage all employees
• Stimulate near-miss reporting
• Apply quality improvement model – Plan-Do-Check-Act
• Build networks
• Challenge accepted practices
• Improve collaboration among management, electrical experts, and safety professionals
• Use standards as tools
• Promote Prevention by Design
• Address life cycle: design, construct, operate, maintain, dismantle

Creating a Continuous Improvement Environment for Electrical Safety. Adapted from Floyd, H.L., et al. (1994).

describe an organizational culture intent on long-term impact on preventing electrical incidents and injuries.

The culture and continuous improvement strategy described in these papers and nurtured for more than 25 years resulted in significant improvement in reducing the severity and frequency of electrical injuries in the company. Most dramatic is the impact on the frequency of fatalities from electrical energy. As shown in Figure 4.1, prior to 1993, fatalities from electrical energy were occurring on average every 33 months. The chart in the figure represents a global workforce of employees and contractors that ranged from 80,000 to 120,000 during this period. Since 1993 and through the submission of this paper in 2012, there have been zero fatalities in company facilities.

The electrical hazards were not reduced, and if anything, the potential for exposure to hazardous electrical energy increased due to the dependence on electrical technologies for energy, control, and communications in industrial applications. What changed was the shift in the electrical safety culture driven by the continuous improvement environment.

The strategy for creating a continuous improvement environment for electrical safety included ten elements discussed below. For more than 25 years, these

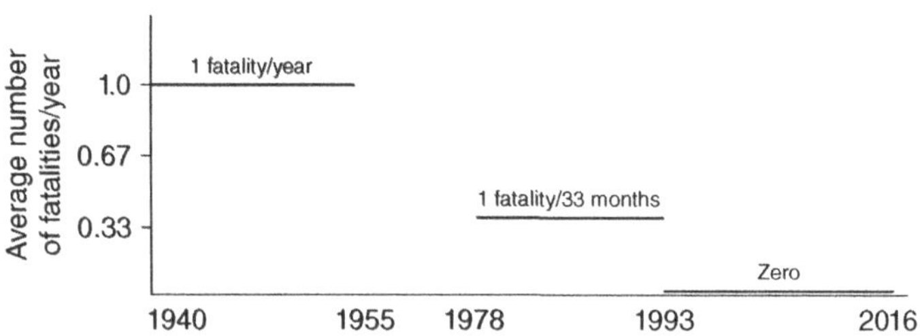

Figure 4.1 Trends in employee and contractor electrocution fatalities in DuPont facilities worldwide.

elements served to guide positive change in electrical safety management in the company.

Understand the Business Consequences

This involves understanding the facts and broad consequences of mishaps in electrical energy and control systems critical to business operations. Nearly every business enterprise is dependent on these electrical systems, and any error, mistake, or improper judgment involving human interaction with these essential electrical systems has several serious potential consequences. There are two primary occupational electrical hazards – electric shock and arc flash burns. Analysis of occupational electrical injury data by Cawley and Brenner shows that electrocution, or fatal electric shock, is the seventh leading cause of occupational fatality in the United States. In addition to the fatal electrocutions, approximately 2,000 workers annually are hospitalized with flash burn injuries, with some injuries permanently disabling or fatal. One in 13 LTIs from electrical contact is fatal, making it one of the most severe injuries that occur in the workplace. Any unplanned disruption of electrical systems has the potential to negatively impact the product quality, raw material utilization, and process safety. Most organizations visibly show commitment to safety for moral reasons – to make sure all workers return home safely at the end of their workday. There are also legal reasons, as government regulations hold employers accountable for providing a safe workplace free from hazards. There is also a direct financial benefit for preventing or reducing the severity of electrical injuries. Compared to common workplace hazards, electrical injuries tend to be more severe, medical costs are higher, and rehabilitation is longer. A study of one utility company by Wyzga and Lindroos found that electrical injuries comprised <2% of total injuries of any given year, yet accounted for 28–52% of total medical costs for the company's workers' injuries. Another study by Liberty Mutual, a major underwriter of workers' compensation insurance, found electrical injuries were the second most expensive workers' compensation claim. This means that the risk of financial loss from electrical injuries is much greater than the low frequency of electrical injuries would tend to indicate. Understanding these facts helps enable better decision-making in allocating resources for preventing electrical injuries.

The lesson: Become a student in understanding the business implications of program success – in the broadest context possible. This will help enable marketing and justification of your program. Preventing incidents in electrical energy, control, and communications systems critical to your business impacts uptime, continuity of operations, energy utilization, workforce development, community image, personnel safety, and process safety. Depending on the business or type of facility, there may be other performance objectives that are impacted. Building a culture that understands these broad implications can help establish a broad foundation of support for this work.

Engage All Employees

Historically, the company's electrical safety program was focused on workers whose job responsibilities specifically included working on or near potentially energized electrical circuits and equipment, i.e. electricians, electrical technicians, and electrical engineers. Fatal injury data showed that this paradigm, for which workers were likely to be injured from electrical hazards, was wrong. Of the 12 fatalities from electrical hazards that occurred in company facilities from 1968 to 2001, 5 were in electrical trade or engineering support functions. Seven were in other work groups or functions, including carpenter, window washer, painter, operations supervisor, and sales and marketing. This experience within the company is aligned with Cawley and Brenner's analysis of US occupational electrical injury and fatality data, showing that electrical trades and electrical support functions accounted for less than ½ of fatalities from contact with electricity. In order to change the paradigm that electrical safety was not just about electrical workers, the entire company was engaged in understanding who was at risk and who could influence electrical safety. This had vertical and horizontal implications for the line organization. Vertically, this effort sought to engage line supervision and management to strengthen their role and involvement in electrical safety. Horizontally, this meant all workers, regardless of their previously perceived exposure to electrical hazards, were considered to have some risk and exposure to electrical hazards. May Electrical Safety Month, introduced in 1990, was the centerpiece of efforts to engage the entire organization. This annual campaign focuses on awareness of electrical hazards in the home as a way to establish understanding that everyone, regardless of job responsibilities, has some risk from electrical hazards (Figure 4.2).

The Lesson: Electrical safety is not just about safety for electrical workers.

Stimulate Near-Miss Reporting

Initially, the database of electrical incidents in the company operations included only those events resulting in shock or burn injuries. This was a very small dataset and was not useful in understanding how best to focus improvement efforts. To stimulate near-miss reporting, the definition for an electrical incident shown in Table 4.2 was developed to help educate workers and management on how to better recognize opportunities to correct the underlying contributing causes to electrical injuries. Part 3 of this definition was intended to capture

Figure 4.2 A poster from the early 1990s campaign promoting May Electrical Safety Month. Note to editor, image concept created by author.

Table 4.2 The definition of an electrical incident that served to increase near-miss (or near-hit) incident reporting and sharing of lessons learned.

Electrical incident: An event resulting from personnel action or equipment failure involving electrical installation, portable electrical equipment, or electrical test equipment that has the potential to result in an injury due to:

1. electrical flash or bum,
2. electrical shock from a source greater than 50 V, or
3. reflex action to an electrical shock.

nonshock injuries, such as falls or lacerations, which were due to intentional or unintentional reaction to a shock sensation.

Within five years, electrical incident reports had increased by more than 10 times, providing valuable insight on where to focus intervention strategies.

The lesson: Electrical injuries are low-frequency/high-consequence events. The lack of, or low number of, electrical injuries in an organization may be due to the inherently low frequency of injury as compared to other hazards in the workplace – and not due to having an effective electrical safety program in place.

Investigating electrical events that do not involve injury can provide invaluable insight into electrical safety program quality and identify systemic opportunities for improvement.

Apply Quality Improvement Model – Plan-Do-Check-Act

This element was intended to establish the expectation that the electrical safety improvement initiative was not a short-term improvement project, but a long-term commitment to continuous improvement. This is aligned with requirements in occupational safety management system standards that emerged beginning in the mid-1990s, including OSHAS 18001 *Occupational Safety and Health Management Systems – Requirements*, ILO-OSH-2001 *Guidelines on Occupational Safety and Health Management Systems*, ANSI Z10-2005 *Occupational Health and Safety Management Systems*, and CSA Z1000-2006 *Occupational Safety and Health Management*. Each of these standards embraced the Plan-Do-Check-Act quality management model, shown in Figure 4.3, attributed to W. Edwards Deming as the means of ensuring long-term sustainable improvement. This model is embedded in essentially every successful business management improvement process, including the Malcolm Baldridge Quality Improvement Criteria and 6 Sigma. It is simple and powerful when executed with discipline and rigor.

The lesson: Safety management system standards provide the framework for helping to ensure long-term management commitment for sustainable improvement.

Build Networks

Internal networks within the company were created to link all people who were involved in or impacted by the electrical work environment in the company.

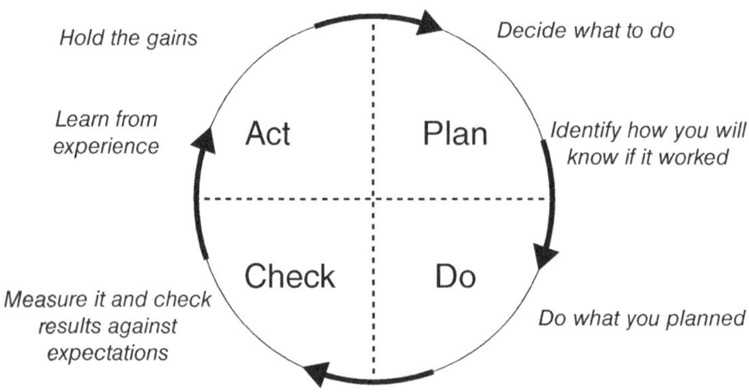

Figure 4.3 The Deming quality improvement model.

This enabled the discovery of best practices and the dispersion of these practices across all locations. External networks, which reached outside the company, were essential in connecting with other centers of excellence and in engaging other companies and organizations in advancing codes, standards, and inherently safer technologies. The IAS Petroleum and Chemical Industry Committee provided an invaluable platform for building and nurturing collaborative networking. The most visible and successful of these external networks is the IEEE IAS Electrical Safety Workshop. This annual forum was launched in 1992 with 35 participants. In recent years, attendance has exceeded 600. It is the longest running international forum dedicated solely to advancing the prevention of occupational electrical injuries and fatalities.

An example of results derived from external networking was the collaborative effort in 1996 involving eight companies to create an awareness of arc flash hazards in industrial settings. This collaboration produced materials that were used to promote awareness and education of arc flash hazards and served as a catalyst to stimulate innovation in prevention and protection technologies.

The lesson: Invest time and energy in networking engagements. It is not enough to be "connected." Being actively engaged in conversation, debate, and collaboration with an expanding network of professional colleagues will stimulate and accelerate possibilities and solutions. Building networks entails creating and nurturing relationships with people from a broad range of experiences and disciplines. Do not limit yourself to maintenance and reliability disciplines. Sometimes, the greatest breakthroughs in a technology are inspired by people outside the particular discipline. Value other perspectives, and look for analogies in other fields of study. Your network may be within your own organization and could extend into different functions. Professional organizations provide valuable opportunities to learn from others who are dealing with the same problems you face, but may be applying different solutions.

Challenge Accepted Practices

Electrical fatalities were not on the company's radar screen because electrical injuries were not significant when looking at the total recordable injury data. One of the tenets of safety management in the 20th century was the Accident Triangle, developed by William Heinrich based on his research in the 1920s. Heinrich popularized the theory that eliminating near-miss incidents and minor injuries will result in a reduction of more severe injuries, including fatalities. As the company focused on preventing severe electrical injuries, one of the first realizations was understanding that efforts to manage safety based on data from total recordable injuries tended to discount the severity of electrical injuries. Recognizing that the Heinrich accident triangle model did not accurately represent electrical incident

MEASURING, TESTING, AND INSTRUMENTS

154. Electricians often test circuits for the presence of voltage by touching the conductors with the fingers (a lethal but common practice in the late 19th and early 20th centuries).

This method is safe where the voltage does not exceed 250 and is often very convenient for locating a blown-out fuse or for ascertaining whether or not a circuit is alive. Some men can endure the electric shock that results without discomfort whereas others cannot. Therefore, the method is not feasible in some cases. Which are the outside wires and which is the neutral wire of a 115/230-volt, three-wire system can be determined in this way by noting the intensity of the shock that results by touching different pairs of wires with fingers. Use the method with caution and be certain that the voltage of the circuit does not exceed 250 before touching the conductors.

Figure 4.4 Excerpt for the seventh edition of *American Electricians Handbook* published in 1953. This practice is now considered unsafe and potentially fatal. Adapted from American Electricians Handbook published in 1953.

and injury statistics was an important discovery. It sets the stage for focusing on an improvement initiative without being based on total injury statistics. This is discussed in more detail in Section 4.3.

Another example involved the development of the concept "Test Before Touch," created to help displace the unsafe practice of intentionally touching conductors to verify the absence of voltage that had evolved in the early days of industrial electrification. Prior to the availability of portable voltage testing instruments, electricians developed techniques for testing for the absence of voltage by touching with bare fingers. This practice was documented in the *American Electricians Handbook* in editions published from 1913 to 1961. An excerpt from the seventh edition published in 1953 is shown in Figure 4.4. This practice is now known to be potentially fatal. To firmly establish the practice of never assuming a circuit is de-energized and to always use an appropriate testing instrument with voltage-rated glove personal protection, the company created the slogan "Test Before Touch" and created supporting education and awareness materials, including posters and monthly newsletters.

The lesson: Challenge what you believe to be true. Maintain a curiosity for seeking a better solution. Breakthrough improvement often involves abandoning old ideas for new ones. Continuous improvement entails continuous learning. New ideas may appear contrary to established practices. Letting go of established practices may be the hard part in exploring new methods and techniques.

Improve Collaboration Among Management, Electrical Experts, and Safety Professionals

Prior to 1990, the electrical safety program had been delegated to electrical experts. In retrospect, the knowledge base of this group was extensive in electrical technologies and work practices, but weak in hazard analysis, risk assessment, and safety management systems. Safety professionals typically did not get involved in the electrical safety program, except during injury investigations. Efforts were taken to create and nurture collaboration among electrical experts, safety professionals, and management to strengthen all aspects of the electrical safety program.

Management has the responsibility for managing priorities, resources, and setting business objectives. Safety professionals bring a skill set in safety management systems, risk management, and rapport with all levels of the line organization. Electrical engineers, electricians, and technicians bring a skill set spanning design, construction, maintenance, and operation of electrical equipment and systems. Collectively, these skills, knowledge, and responsibilities helped create an extraordinary collaboration and synergy to assess and improve the company's electrical safety program.

The lesson: Do not assume electrical experts are experts in safety management. Engage safety professionals. Do not assume management has all the information to make the best decisions in support of your program. Management's role is to set goals, prioritize objectives, allocate resources, and assess progress. When there is incomplete knowledge and data, optimum decisions will likely not result. Your job includes educating management to enable the smartest decisions possible to support your program.

Use Standards as Tools

An organization may choose compliance with codes, standards, and regulations as a goal. However, compliance may achieve legal requirements without achieving world-class performance. Industry consensus standards, by definition, imply compromise by members of the developing organization. Leading-edge, state-of-the-art, or innovative methods may take decades to be incorporated into consensus standards. The company took the approach that compliance was a step in the process to achieve electrical safety performance objectives. Looking beyond compliance, electrical safety leaders nurtured a commitment to seek collaboration with other companies and organizations to promote and stimulate innovation in system and equipment design, safe work practices, tools and equipment, and training methods.

The lesson: Industry consensus standards may be historical documents of 20-year-old innovative ideas. To be truly state of the art, an organization may need to look beyond standards and be curious of what other leading organizations are doing. An effective electrical safety program goes beyond compliance and is built on guidance and requirements from a wide resources base of electrical safety-specific codes, standards, regulations, and recognized best practices that are integrated within the framework of a safety management system.

Promote PtD

Initial efforts in arc flash hazard mitigation focused on protective measures, primarily thermal protective clothing and PPE. Company engineers soon began exploring engineering design techniques to reduce thermal energy exposure in arc flash events. An early success was realized by retrofitting fast-acting current-limiting fuses in 480 V feeder circuits. This paved the way for further application of engineering design solutions to prevent or reduce the severity of exposure, including arc-resistant designs, advanced protective relaying schemes, touch-safe terminal designs, remote switching and operation, and power distribution and motor control equipment that enabled troubleshooting via "smart" technologies.

The lesson: Recent developments, in particular the NIOSH Prevention through Design initiative and the publication of ANSI Z590.3 *Prevention through Design Guidelines for Addressing Occupational Hazards and Risks in Design and Redesign Processes*, confirm that there is much room for improvement in the concepts of eliminating or reducing hazards in the design of facilities, equipment, and processes.

Address Life Cycle: Design, Construct, Operate, Maintain, Dismantle

This is about taking a long-term, comprehensive view of electrical safety, from the initial design concept all the way to demolition or replacement of the facility. All stages in an electrical system's life cycle affect the electrical work environment. An analysis of electrical incidents and injuries in the company showed that design choices biased toward inherently safer design could help prevent electrical injuries, not only in the operate/maintain phase of the facility but also during construction, commissioning, and end-of-life dismantlement.

The lesson: Building codes typically address safety of personnel who may be exposed to hazards during the operate/maintain phase of a facility. Decisions made in the design and construction planning can impact the risk of electrical injury for the life of the facility, including construction and demolition.

A Shift in Safety Management Mindset

The company's efforts to change the electrical safety culture in its operations parallel a more fundamental change in safety management theory to more effectively manage occupational hazards that share the characteristic of low frequency, when compared to other occupational hazards, but have high consequence, including the likelihood of serious disabling injury or fatality.

Low-Frequency/High-Consequence Injuries

The performance improvement in reducing the frequency of electrical fatalities provides a case study in an emerging field of advanced safety management of low-frequency/high-consequence hazards. There is a subset of hazards in the workplace that contribute relatively few injuries, but a much higher percentage of fatalities and serious disabling injuries. Over the past 30 years, thought leaders in safety management have challenged traditional methods in injury prevention as being ineffective in addressing hazards that are associated with injuries that occur relatively infrequently, but have extraordinary severity potential. As theories in advanced safety management, with an emphasis on fatality prevention, have evolved, the case history in electrical safety improvement validates the concerns and possibilities of the management's focus on low-frequency/high-consequence injuries.

The Heinrich Accident Triangle

The accident triangle, shown in Figure 4.5, developed by Heinrich in the 1920s, has served as a model for safety management for nearly 80 years. This model implies a fixed ratio of near-miss incidents to minor injuries to major injuries. Although this has been a valuable model, Johnson noted its limitation in that it is based on homogeneous treatment of all hazards and does not acknowledge a wide range of severity potential and a wide range of ratios between minor injuries and major injuries. As early as 1980, Peterson and Roos began questioning aspects of this model, in particular, the application of the ratios to all hazards. More recent work

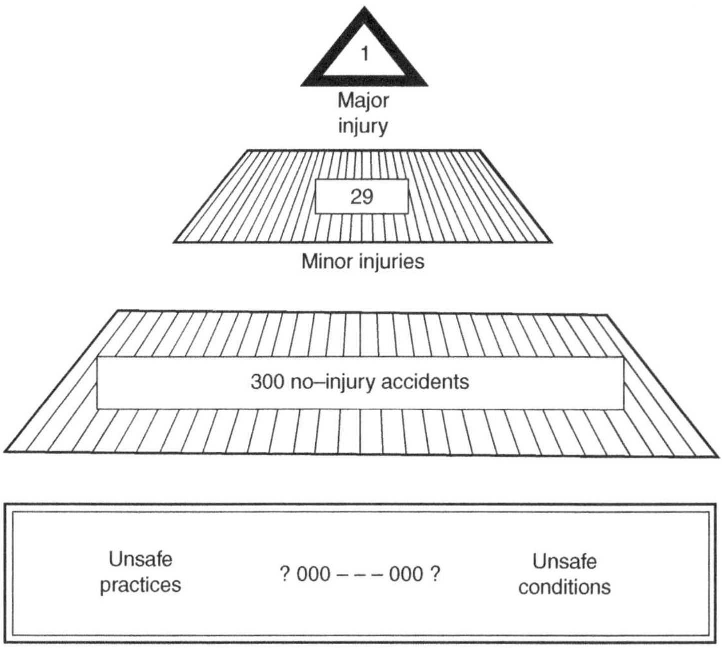

Figure 4.5 Accident triangle first published by H.W. Heinrich in 1931.

by Manuel and Krause proposed that hazards having high potential for severe injuries are not accurately described with this model and need to be addressed differently than those with low potential for serious injury. Manuel and Krause have also noted that focusing on reducing the frequency of injuries may show successful results without a corresponding reduction in severe injuries. Anderson and Denkl conducted an analysis of 2007 US Bureau of Labor Statistics data on 5,657 fatalities and 1,158,870 LTIs and concluded that Heinrich's model does not correlate to injuries and fatalities from hazards that have high potential for fatality.

To illustrate the limitation of applying the accident triangle to electrical safety, consider the question: How many minor incidents involving 12-V automotive

Table 4.3 Occupational fatalities by cause in the United States, 2010.

Type of Fatality	Number of Fatalities (2010)
Total	4, 547
Transportation. *Excludes water, rail, air*	1, 519
Assaults and violent acts	808
Falls	635
Struck by object or equipment	402
Caught in or compressed by equipment	224
Exposure to harmful substance or environment	246
Contact with electric current	**163 (3.58%)**
Aircraft	151
Caught in or crushed by collapsing materials	91
Water vehicle	52
Explosions	78
Railway	44
Other	134

Extracted from data compiled by Electrical Safety Foundation International from US BLS data.

batteries must be eliminated to prevent an electrocution fatality at the top of the triangle? Clearly, the answer is that there is no relationship between these two. The intervention activities at the bottom of the accident triangle need to be hazard specific.

Tables 4.3 and 4.4 help illustrate a flaw in Heinrich's model. The data in these figures shows that the hazards that are likely to cause fatality produce a relatively small percentage of nonfatal injuries. Contact with electricity causes just 0.16% of all recordable injuries, but 3.58% of all fatal injuries, and illustrates the concept of low frequency/high consequence.

Table 4.4 is based on analysis by Cawley and Brenner of nonfatal occupational injuries in the United States.

Table 4.4 shows the data in tabular form. Floyd and Floyd adapted Table 4.5 from studies by Anderson and Denkl and Cawley and Brenner. This figure shows the ratio of LTIs to fatalities by the type of incident. The severity index ranges from 12 for fires and explosions to 14,033 for overexertion in lifting. In other words, 1 in 14,034 lifting injuries is fatal, but 1 in 12 fire and explosion injuries is fatal. For electrical injuries, 1 in 13 is fatal. This chart clearly shows that assuming a fixed ratio of minor injuries to fatalities, as was the basis for Heinrich's model, can be misleading in how an organization perceives the risk of injury. Manuel and Krause have also noted that focusing on reducing the frequency of injuries may show successful results without a corresponding reduction in severe injuries.

Table 4.4 Comparison of select nonfatal occupational injuries in the United States, 2010.

Type of Nonfatal Injury	Number of Injuries (2010)
Total	1, 191, 100
Sprains, strains, tears	474, 000
Musculoskeletal disorders	346, 300
Falls on the same level	182, 400
Struck by object	138, 530
Falls to lower level	73, 520
Assault/violent act by person	40, 310
Highway accidents	36, 460
Assault/violent act by animal	7, 160
Fires and explosions	3, 000
Electrical shock and burn	**1,890 (0.16%)**

Adapted from U.S. BLS Economic News Release (2010).

Table 4.5 Data from US Bureau of Labor Statistics showing the ratio of lost-time injuries to fatalities.

Event or Exposure	LTI/Fatality Ratio[a)]
Fires and explosions	12
Contact with electricity[b)]	**13**
Transportation accidents	23
Assaults and violent acts	28
Fall to a lower level	104
Exposure to harmful substance or environment	107
Caught in, compressed or crushed	134
Struck by object	323
Falls on same level	2,056
Struck against object	8,414
Slips or trips without fall	12,593
Overexertion in lifting	14,033

a) Lost-time injury to fatality ratio.
b) Contact with electricity data from Cawley and Brenner; original data from Anderson and Denkl.

Adapted by Floyd and Floyd (2014) from Anderson and Denkl with electrical injury data from Cawley & Brenner.

As shown in Table 4.5, the potential for an electrical injury to be fatal, as indicated by the LTI/fatality ratio, is one of the highest of all hazards in the workplace. For electrical safety (as well as other low-frequency/high-consequence risks), *the driving force for prevention has to focus on consequence severity rather than injury frequency*.

Accidents that are low in frequency can also mean that a company or organization does not have sufficient internal data on electrical injuries to be statistically meaningful. Due to the inherent low-frequency occurrence of electrical injuries, an individual or company may not recognize the potential for a fatal (high-consequence) injury.

The lack of, or low number of, electrical injuries is not a valid indicator of the quality of the electrical safety program. Factors that may be better indicators of electrical safety program quality would be:

- Frequency and quality of electrical safety training for electrical workers, non-electrical workers, and line supervision
- Frequency and quality of field audits that examine the implementation of the organization's electrical safety program

- Frequency and quality of management systems audits focused on preventing exposure to electrical hazards
- Attention to inherently safer design in the selection of hardware and electrical system design in capital projects
- Discipline in maintaining maintenance programs for equipment and systems critical to electrical safety

A New Model – The Safety Risk Pyramid

The evolution in advanced safety management has brought new thinking in how to manage both the severity and frequency of injuries from workplace hazards. Figure 4.6 is an updated model, based on Heinrich's accident triangle, but it brings in advanced concepts in safety management. It includes the notion that certain hazards have a much smaller base of minor injuries with respect to fatalities, and that mitigation programs must be hazard specific. This model can be helpful in assessing whether specific hazards are being appropriately addressed and should certainly be applied to electrical safety.

Heinrich's model did not differentiate hazards by severity but was based on a simplistic but logical concept that by reducing unsafe acts and conditions, minor and serious injuries could be reduced proportionally. However, this model does

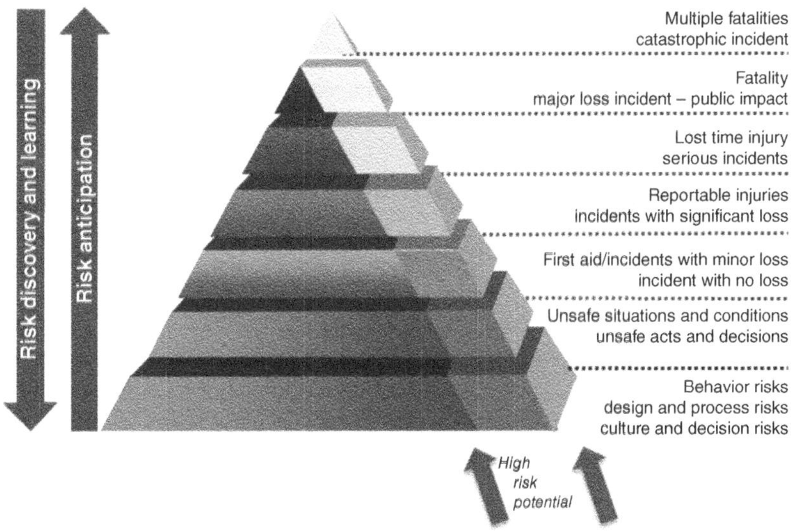

Figure 4.6 © used with permission. A safety risk pyramid, an updated model of the Heinrich accident triangle that incorporates the concept of hazards with a higher severity index. Lin and Solomon [1].

not acknowledge that some hazards are significantly more likely to cause serious injury and fatality than others, making it misleading regarding hazards that have more severe consequences.

Fundamental to understanding risks of electrical injury is awareness of the Low-Frequency/High-Consequence nature of electrical injuries. As noted earlier, electrical injuries account for a small number of nonfatal injuries (0.16% in 2010, see Table 4.4) and a small but larger number of fatalities (3.58%, see Table 4.3). Along with the high potential for fatality is the high consequence of financial costs associated with electrical injuries. This means that the potential magnitude of financial loss from electrical injuries is much greater than the low frequency of electrical injuries would tend to indicate.

Taken together, Tables 4.3, 4.4, and 4.5 show that assuming a fixed ratio of minor injuries to fatalities can be misleading in how an organization perceives the likelihood of serious injury or fatality. This may contribute to an organization being vigilant about electrical safety *immediately following an injury*, but lapsing in their practice after a certain amount of time has passed without an incident.

Over 20 years of experience with implementing the strategy for creating a continuous improvement environment for electrical safety presented in 1992 has demonstrated measurable and sustainable results and shows that significant improvement in electrical fatalities is possible. Recent advances in safety management focusing on low-frequency/high-consequence injuries are validating the strategy for creating a continuous improvement environment while presenting new concepts and tools for enhancing the strategy. This case study provides insight on how other occupational hazards characterized as low frequency/high consequence could be more effectively managed to prevent serious disabling injuries and fatalities.

Crosstalk: Questions for Discussion

1. Do you know the history of electrical injury performance for your organization?
2. What can you do to increase reporting and learning from near-miss incidents?
3. Does your organization use the concept of May Electrical Safety Month, sponsored by the Electrical Safety Foundation?
4. What can you do to increase the awareness of low-frequency/high-consequence events?
5. What are you most proud of in your electrical safety program? Why?

Reference

1 Lin, M.-L. and Solomon, P. (2012). Tales of the unexpected. *Safety and Health Practitioner* 37–40.

Further Reading

Anderson, M. and Denkl, M. The Heinrich Triangle – too simplistic a model for HSE Management in the 21st century? *Society of Petroleum Engineers 2010 International Conference on Health Safety and Environment.*

ANSI Z590.3. Prevention through design guidelines for addressing occupational hazards and risks in design and redesign processes, American Society of Safety Engineers, Des Plaines, Illinois

ANSI/AIHA Z10-2005. Occupational health and safety management systems, American Industrial Hygiene Association.

Bonner, S., Eastwood, K., Floyd, H.L. et al. (2002). Key learnings from the IAS Electrical Safety Workshop. *IEEE Industry Applications Magazine* 8 (3): 11–17.

Capelli-Schellpfeffer, M., Downey, R.E., Jamil, S. et al. (2000). *Staged Tests Increase Awareness of Arc-Flash Hazards in Electrical Equipment.* IEEE Transactions on Industry Applications.

Cawley, J.C. and Brenner, B.C. (2013). Occupational electrical injuries in the U.S., 2003–2009. *IEEE Industry Applications Magazine* 19 (3): 16–20.

Crawford, C. and Haggerty, N.K. (2008). Test before touch – easier said than done. *IEEE Industry Applications Magazine* 14 (3): 32–40.

CSA Z1000-2006. Occupational safety and health management, Canadian Standards Association, Mississauga, ON.

Doughty, R.L., Epperly, R.A., and Jones, R.A. (1992). Maintaining safe electrical work practices in a competitive environment. *IEEE Transactions on Industry Applications* 28 (1): 196–204.

Floyd, H.L. and Cole, B.C. (2015). Twenty years later: creating a continuous improvement environment for electrical safety. *IEEE Transactions on Industry Applications* 51 (3): 2633–2640.

Floyd, A.H.L. and Floyd, H.L. (2014). Cultural drift and the occlusion of electrical safety. *IEEE Transactions on Industry Applications* (May/June 2014).

Floyd, H.L., Cole, B.C., Doughty, R.L. et al. (1994). Creating a continuous improvement environment for electrical safety. *IEEE Transactions on Industry Applications* 30 (3): 543–552.

Heinrich, H.W., Peterson, D., and Roos, N. (1980). *Industrial Accident Prevention – A Safety Management Approach.* New York: McGraw Hill.

ILO-OSH 2001. Guidelines on occupational safety and health management systems, International Labour Organization, Geneva.

Johnson, A. Were Herbert William Heinrich's theories valid, and do they still matter? National Safety Council, retrieved at www.nsc.org/safetyhealth/Pages/Examiningthefoundation1011.aspx#.UXVDJ7Up_gd ().

Krause, T.R. (2010). A new paradigm for fatality prevention, *Safety 2010 Conference*, American Society of Safety Engineers (13–15 June 2010), Baltimore, Maryland.

Work related electrical injuries: study sparks new insights, *Liberty Mutual Research Institute for Safety*, vol 13, No. 3, 2010.

Lippincott, W.F. (1954). Is "hot". *Electrical Work Necessary? DuPont Safety Conference*, Chicago, IL.

Manuele, F.A. (2002). *Heinrich Revisited: Truisms or Myths*. Itasca, NY: National Safety Council Press.

Manuele, F.A. (2004). Injury ratios. *Professional Safety* 49 (2): 22–30.

Manuele, F.A. (2008). "*Serious Injuries & Fatalities – A Call for a New Focus on Their Prevention*", *Journal of Professional Safety*, 32–39. American Society of Safety Engineers.

OHSAS-18001. Occupational safety and health management systems – requirements, The Occupational Health & Safety Group, Cheshire, UK.

Wyzga, R.E. and Lindroos, W. (1999). Health implications of global electrification. *Annals of the New York Academy of Sciences* 888: 1–7.

5

Our Electrical World

Life in 1900 was a constant struggle against disease, starvation, pollution, treacherous working conditions, and information sharing delays with enormous cultural divides unbreachable with communications technologies of the day. By the end of the 20th century, the world had become a healthier, safer, and more productive place, primarily because of engineering achievements.

On February 22, 2000, astronaut/engineer Neil Armstrong, speaking on behalf of the National Academy of Engineering, announced the 20 engineering achievements that have had the greatest impact on quality of life in the 20th century. The announcement was made during National Engineers Week 2000 at a National Press Club luncheon. The achievements – nominated by 29 professional engineering societies – were selected and ranked by a distinguished panel of the nation's top engineers. The panel, chaired by H. Guyford Stever, former director of the National Science Foundation (1972–1976) and Science Advisor to the President (1973–1976), worked in anonymity to ensure the unbiased nature of its deliberations.

"As we look at engineering breakthroughs selected by the National Academy of Engineering, we can see that if any one of them were removed, our world would be a very different – and much less hospitable place," said Armstrong. "Almost every part of our lives underwent profound changes during the past 100 years, thanks to the efforts of engineers, changes impossible to imagine a century ago. People living in the early 1900s would be amazed at the advancements wrought by engineers," he said, adding, "as someone who has experienced firsthand one of engineering's most incredible advancements – space exploration – I have no doubt that the next 100 years will be even more amazing."

The top engineering achievement was Electrification – the vast networks of electricity that power the developed world. Electrification powers almost every pursuit and enterprise in modern society. It has literally lit the world and impacted

Hidden Risk in Occupational Electrical Safety: Through the Lens of System Safety,
First Edition. H. Landis "Lanny" Floyd.

countless areas of daily life, enabling every one of the other 19 achievements listed as follows:

1) **Automobile:** Revolutionary manufacturing practices made the automobile the world's major mode of transportation by making cars more reliable and affordable to the masses.
2) **Airplane:** Flying made the world accessible, spurring globalization on a grand scale.
3) **Safe and abundant water:** Preventing the spread of disease, increasing life expectancy.
4) **Electronics:** Vacuum tubes and, later, transistors that underlie nearly all modern life.
5) **Radio and television:** Dramatically changed the way the world received information and entertainment.
6) **Agricultural mechanization:** Leading to a vastly larger, safer, less costly food supply.
7) **Computers:** The heart of the numerous operations and systems that impact our lives.
8) **Telephone:** Changing the way the world communicates personally and in business.
9) **Air conditioning and refrigeration:** Beyond convenience, it extends the shelf life of food and medicines, protects electronics, and plays an important role in health care delivery.
10) **Interstate highways:** 44,000 miles of US highway allowing goods distribution and personal access.
11) **Space exploration:** Going to outer space vastly expanded humanity's horizons and introduced 60,000 new products on Earth.
12) **Internet:** A global communications and information system of unparalleled access.
13) **Imaging technologies:** Revolutionized medical diagnostics.
14) **Household appliances:** Eliminated strenuous, laborious tasks, especially for women.
15) **Health technologies:** Mass production of antibiotics and artificial implants led to vast health improvements.
16) **Petroleum and gas technologies:** The fuels that energized the 20th century.
17) **Laser and fiber optics:** Applications are wide and varied, including almost simultaneous worldwide communications, noninvasive surgery, and point-of-sale scanners.
18) **Nuclear technologies:** From splitting the atom, we gained a new source of electric power.
19) **High-performance materials:** Higher quality, lighter, stronger, and more adaptable.

We live in an electrical world, with nearly every aspect of modern business and commerce dependent on electrical technologies and interactions with tools, appliances, equipment, and systems having some aspect of potential exposure to inherently hazardous electrical energy. Although the degree of risks ranges from little to great, it is the norm today that all people have some risk of injury from electrical energy, whether at work, at home, or going about their daily lives in public places. The Electrical Safety Foundation was established in 1994 with the sole purpose of promoting awareness and education of electrical hazards and risk of injury. The Foundation is a source of factual information and materials, and it sponsors May Electrical Safety Month as an annual campaign to promote its mission. Occupational electrical injury and fatality data available from the US Bureau of Labor Statistics is used to produce an annual report posted on the ESFI.org website. Examples of past reports and analyses are included in Appendix of this book.

All Workers Have Risk for Exposure

For many organizations, the focus and improvement in occupational electrical safety programs have largely addressed electrical workers whose primary work tasks involve the construction, operation, and/or maintenance of electrical equipment. However, nearly 50% of workplace electrical injuries and fatalities are not electrical workers. Managers and administrators, painters, truck drivers, farm workers, groundskeepers, and gardeners are among the top 10 occupations having the most fatal electrical injuries. For these nonelectrical workers, or "other" workers, the exposure to electrical hazards ranges from the use of common portable tools, extension cords, and appliances to unintentional contact with overhead power lines in the course of routine work activities. Workplace electrical safety programs tend to focus on safeguarding workers whose job activities involve working on or near electrical equipment and circuits that either are or potentially could be energized. Workers whose exposure to electrical hazards is generally limited to unintentional contact with unguarded or exposed conductors may not be specifically addressed or covered by the program. A literal interpretation of the scope statement in NFPA 70E Article 90.2 could unintentionally limit the application of the useful guidance in this standard. The reality is that we live in an electrical world, with nearly every aspect of modern business and commerce dependent on electrical technologies and interactions with tools, appliances, equipment, and systems that have inherently hazardous electrical energy. Although the degree of risks ranges from little to great, it is the norm today that all people in the workplace have some risk of injury from electrical energy. Equipment, appliance and tool designs, and installation requirements such as those found in NFPA 70E, *National Electrical Code*, and IEEE/ANSI C2,

National Electrical Safety Code, have continually evolved to reduce the risk of electrical injury from normal interaction. When the inherent safety features of electrical equipment or systems are not as intended due to damage, wear, improper installation, misapplication, misuse, or the hazard being unrecognized, the risk of electrical injury to the unsuspecting worker can significantly increase. Examples include a mobile television news crew that parks their vehicle underneath a power line and raises an antenna mast into the line, a carpenter who uses a power tool plugged into an extension cord with a damaged ground wire, a roofer handling a section of aluminum gutter near an overhead electric line, or a cover on an electric powered appliance is missing or unsecured creating a booby trap for an unsuspecting worker.

Implications

The misperception of who may be at risk for electrical injuries may be a limiting factor in achieving the potential for injury prevention and the application of best practices for preventing these injuries. Analysis of BLS data by Cawley and Brenner showed that 31% of electrical fatalities occurred in occupations that may not be considered as having a high risk of electrical injuries. These occupations included construction laborers, managers, truck drivers, agricultural workers, roofers, painters, and landscaping and grounds-keeping workers. Taylor et al. noted that managers and administrators, painters, truck drivers, farm workers, groundskeepers, and gardeners are among the top 10 occupations having the most fatal electrical injuries. Data from the Electrical Safety Authority of Ontario shows that 71% of occupational electrocution fatalities for the period 2001–2010 were workers in nonelectrical trades.

It is essential that workplace electrical safety programs address electrical safety for all workers and not just workers whose job responsibilities involve working on or near energized electrical circuits. For these nonelectrical workers, the exposure to electrical hazards ranges from the use of common portable tools and appliances to unintentional contact with overhead power lines in the course of routine work activities.

This raises the questions: What is the workplace? What workers are at risk of electrical injury? Do existing electrical safety programs encompass all workplaces and all workers? Article 90.1 of NFPA 70E states, "The purpose of this standard is to provide a practical safe working area for employees relative to the hazards arising from the use of electricity." The scope description in Article 90.2 further narrows the application of the standard with this statement: "This standard addresses electrical safety requirements … for the practical safeguarding of employees during activities such as the installation, operation, maintenance, and demolition of electric conductors, electric equipment, signaling and communications conductors and equipment, and raceways." Strict adherence to the boundaries in these statements could limit the application of this standard to workers whose job description specifically requires potential exposure to energized electrical equipment and systems. Could any of the requirements in the standard apply to other workers? What could be the impact if there were more effort to facilitate the application to other workers who have potential exposure to electrical hazards in their work activities?

DuPont, the company described in this case history in Chapter 4 is among the oldest of the Fortune 500, having been incorporated in 1802. For the first 100 years, the company produced explosives used in farming, construction, hunting, and warfare. The manufacture and distribution of these products were inherently hazardous and fostered a robust and disciplined culture for managing safety for highly hazardous processes. At the end of the 19th century, electricity was introduced in the company's facilities, and the experience in managing safety was applied to this new technology. Over a 33-year period 1968–2001, employees and contractors in its global operations incurred 12 fatalities from electrical hazards. The combined employee and contractor workforce during this period ranged from ~80,000 to ~120,000. The "workplace" included more than 200 industrial facilities, refineries, construction sites, laboratories, research centers, retail outlets, farming operations, and office complexes throughout the Americas, Asia, and Europe. Of the 12 fatalities, 1 was a supervisor standing on a coal pile underneath an overhead power line, 3 were electricians in separate incidents working on equipment thought to be de-energized, 2 were carpenters who rolled a scaffold into an overhead electric line, 1 was a window washer who contacted an overhead power line while positioning a ladder against a building, 1 was a welder who attempted to troubleshoot a welding machine, 1 was a painter who came in contact with a high-voltage terminal on top of a substation transformer, 1 was a supervisor providing troubleshooting assistance, 1 was an engineering consultant performing a visual inspection of electrical switchgear, and 1 was a sales representative who came in contact with an energized conductor in a customer's agricultural operations. As summarized in Table 5.1, of the 12 fatalities, 3 were electricians, and only 5 could be characterized as workers whose job responsibilities would routinely place them near exposed electrical hazards.

Data from Taylor et al. and Cawley and Brenner shows that while electricians, electrical apprentices, and electrical power installers are the highest at-risk groups, approximately half of electrical fatalities fall outside these electrical trades. Construction laborers, groundskeepers and gardeners, truck drivers, carpenters, managers and administrators, and painters are among the top 10 occupations sustaining the most fatal electrical injuries. Lombardi et al. have provided insight on work activities contributing to no fatal electrical injuries. Manual tasks

Table 5.1 Job description of workers suffering fatal electrical injuries in DuPont, 1968–2001.

• Carpenter (2) • Electrician (3) • Electrical craft supervisor • Electrical consultant • Painter • Powerhouse coal operations supervisor • Sales and marketing representative • Welder • Window washer

characterized as plugging/unplugging, loading/unloading, cleaning/dusting, and tasks such as testing/checking power and turning on/off comprise 33.5% of activities contributing to nonfatal electrical injuries. The study by Taylor et al. on fatal electrical injuries shows that vehicular and transportation operations, use of tools and machinery, and materials handling operations accounted for 31.3% of fatal electrical injuries during the period 1992–1999.

Cawley and Homce analyzed occupational electrocution fatalities by event. Contact with overhead lines was the leading cause of occupational electrocution. The second leading cause was contact with wiring, transformers, or other electrical components. This exposure is typical of work activities associated with electrical crafts. The third leading cause is contact with electric current when using machines, tools, or appliances. Everyone, whether at work, at home, or in public places, has potential exposure to extension cords and cord-powered tools and appliances.

Exposure Scenarios

Overhead electric line contact is of particular concern, accounting for 41.4% of electrical fatalities and 4.3% of nonfatal electrical injuries. Contact with overhead lines during activity such as electrical system installation and maintenance is the largest category of workplace electrocutions in the United States. Within this group, contact with a handheld object accounted for 22%, and contact indirectly through a piece of high-reaching mobile equipment, such as cranes, drilling rigs, mobile work platforms, antennas, irrigation rigs, and raised truck beds, accounted for 17%. Anything that can go up in the air has the potential to make unintentional contact with an overhead power line. This includes antenna masts on mobile equipment, water drilling rigs, crane booms, elevated work platforms, ladders, pipes, roof gutters, and building siding. Literally, any conductive object, material, or equipment that can be held or in contact with a person has the potential for extending the individual's reach to unintentionally contact an overhead electric line.

The use of GFCIs has proven effective in significantly reducing fatalities associated with portable tools and appliances. Wells reported more than a 70% decrease in electrocutions involving portable tools and appliances in the 32-year period from 1970 to 2002. Requirements for GFCI installation were introduced in the 1968 National Electrical Code. Each edition since then has expanded the requirements for GFCI installation, and there was an estimated 33 million units installed by 2002. The expanded application of GFCIs was likely a major contributor to the significant reduction in electrocution fatalities. Studies suggest there is much room to expand the use of these devices. A 2001 study by the National Electrical Manufacturers Association discovered that 14% of ground fault CBs and 8% of receptacle GFCIs were nonoperational.

An organization intent to reduce the risk of electrical injury in its operations should consider actions in at least two areas: expand the use of GFCI protection and assess practices and procedures to prevent unintentional contact with overhead lines.

A subtle but very important revision was made in the 2012 edition of NFPA 70E. This phase was added to Article 90.2 Scope: *This standard also includes safe work practices for employees performing other work activities that can expose them to electrical hazards* .... For the first time in its 30+ year existence, NFPA 70E suggested that the standard may be useful in safeguarding workers outside the electrical crafts. This was a subtle revision, but it has the potential to be one of the most significant changes ever made to this standard. This change presents an opportunity to critique our collective actions in supporting the implementation of electrical safety programs to include all who may be exposed to hazardous electricity in our electrical world.

Assessing What Workers Are Included in Your Program

For individuals and organizations that may have perceived themselves as not being at risk of electrical injuries, electrical safety codes and standards may be a bit intimidating. For example, the owner of a small painting contractor firm that does not have an electrical expert on staff could have difficulty in interpreting and applying the requirements of NFPA 70E to employees. With the major emphasis of most electrical safety codes and regulations being directed to subject matter experts or to workers routinely working on or near energized electrical conductors, tailoring electrical craft-specific requirements to apply to nonelectrical work environments may take some effort. Companies with established electrical safety programs are best positioned to critique and expand their program to more effectively include workers in nonelectrical crafts. Examples of questions in such a self-critique could include the following:

- Do we have a documented electrical safety program?
- Is it best characterized as an electrical safety program for workers in electrical crafts, or does it address the risk of electrical injury for all workers?
- Do workers use mobile equipment or handle material that could come in contact with overhead electric lines? How does our program control the movement of mobile equipment and material handling near overhead electric lines?
- Do workers use portable tools, appliances, and extension cords? How do we know GFCIs are being used?
- Do workers use portable tools and appliances in potentially damp or wet environments? How do we know GFCIs are being used?
- Do we have portable tools or appliances operating above 120 V that do not have GFCI protection? Are we applying the latest technology in GFCI protection?
- How disciplined is your the organization in maintaining the integrity of enclosures and covers on electrical equipment, tools, and appliances?
- Are nonelectrical personnel included in arc flash protection programs? Do we understand who is at risk?

These are a few examples of questions that could help identify opportunities to reduce the risk of electrical injuries to all workers. The questions could differ based on the industry, work environment, or maturity of the organization's electrical safety program. There is a common question that all involved in changing and advancing the electrical safety culture should ask. *Do your words and actions narrow or broaden the scope of workers included in an organization's electrical safety program?* Whether you are responsible for an employer's electrical safety program, involved in standards development, provide training in workplace electrical safety, design electrical products or systems, provide products or services in the electrical safety market, or contribute to workplace safety in other ways, your words and actions help establish the culture impacting electrical safety for "other" workers. Taking the time to critically review documents under your control and your own behaviors as projected in spoken words and personal actions may uncover opportunities to make a more positive impact on electrical safety for all workers. Exposure to electrical hazards is a leading cause of occupational fatalities. Consensus standards relevant to electrical safety have evolved in content and application to more effectively reduce the risks from electrical injury. There are opportunities to extend the application of concepts and practices in these standards to workers outside the electrical crafts. This is essential to reduce the total number of occupational fatalities and injuries from electrical hazards, since workers outside the electrical crafts suffer a large percentage of all workplace electrical fatalities and injuries.

Crosstalk: Questions for Discussion

1 How is your work environment dependent on electrical technologies?

2 Looking around you, where does electrical energy exist?

3 What engineering controls are in place to protect you from exposure to hazardous electrical energy?

4 Why is it easy to not think about the electrical hazards that exist in your own home?

5 What does the slogan "Take Electricity Seriously" mean to you?

Further Reading

Brenner, B., Reinvigorating electrical safety, *2011 IEEE IAS Electrical Safety Workshop*, January 30–February 4, 2011, Toronto, ON.

Cawley, J.C. and Brenner, B.C., Occupational electrical injuries in the U.S., 2003–2009, *2012 IEEE IAS Electrical Safety Workshop, January 31–February 3, 2012*, Daytona Beach, FL.

Cawley, J.C. and Homce, G.T. (2008). Trends in electrical injuries in the U.S., 1992–2002. *IEEE Transactions on Industry Applications* 44 (4): 962–972.

Electrical Safety Authority, 2010 Ontario Electrical Safety Report, 400 Shelden Drive, Unit #1, Cambridge, Ontario N1T 2H9, p. 21.

GFCI Field Test Survey, National Electrical Manufacturers Association, January 2001, retrieved from the internet on December 5, 2012 at www.mikeholt.com/htmlnews/afci/GFCInema.pdf.

Green, M.A., Analysis of GFCI data, *U.S. Consumer Product Safety Commission*, May 5, 2004, retrieved from the internet on December 5, 2012 at www.cpsc.gov/volstd/gfci/analysisgfci.pdf.

Lombardi, D.A., Matz, S., Brennan, M.J. et al. (2010). Etiology of work-related electrical injuries: A narrative analysis of workers' compensation claims. *Journal of Occupational and Environmental Hygiene* 6 (10): 612–623.

NFPA 70E (2012). *Standard for Electrical Safety in the Workplace*. Batterymarch, MA: National Fire Protection Association.

Taylor, A.J., McGwin, G. Jr., Valent, F., and Rue, L.W. III, (2002). Fatal electrocutions in the United States. *Injury Prevention* 8 (4): 306–312.

Top Engineering Achievements of the 20th Century, https://www.newswise.com/articles/top-engineering-achievements-of-20th-century.

Wells, J., GFCIs – thirty years and counting, *2003 IEEE IAS Electrical Safety Workshop*, February 4–7, 2003, Houston, TX.

6

Collateral Injuries, Consequences, and Benefits

The Underlying Power of an Effective Electrical Safety Program

Introduction

Modern industry and commerce are nearly completely dependent on electrical technologies for energy, control, data management, and communications. While a worker's mishap in these critical systems has the implication of injury to the worker, with much greater frequency, a mishap or incident includes the consequence of significant disruption to the enterprise's operations.

Operational Excellence

Many businesses and organizations today are pursuing performance goals that are defined in terms of operational excellence. In a broad sense, operational excellence is about improving business and operating processes that enhance product or service quality, achieve higher yields, use less energy, result in less waste, and increase uptime. Most organizations pursuing operational excellence understand that improving safety performance is also good for business. In addition to holding in high regard the moral and humanitarian duty to provide a safe workplace, most organizations regard safety performance as an integral element of business success. An effective electrical safety program has the potential to provide even more benefits for an organization pursuing operational excellence.

Beginning in the late 19th century and continuing today, electrification, or the displacement of other technologies with electrical technology, has transformed all aspects of our modern society. Energy generation and distribution, heating, lighting, transportation, food production and storage, medical diagnostics, and financial transactions are just a few examples in which electrical technology has displaced previously used technologies to enable extraordinary advancements and conveniences. In celebrating the beginning of the 21st century, the National Academy of Engineering identified electrification as the most significant

Hidden Risk in Occupational Electrical Safety: Through the Lens of System Safety,
First Edition. H. Landis "Lanny" Floyd.

engineering achievement of the 20th century. The implications of having such dependence on electrical technology, combined with increasing initiatives in pursuit of operational excellence, make the elimination of errors and failures in electrical systems of paramount importance. Today, it is difficult to imagine our world without the convenience and productivity made possible by electricity and the application of electrical technology. Virtually every aspect of our society, including transportation, leisure activities, business and commerce, food production and distribution, and medical services, is dependent on electricity for energy and control. When mishaps occur in critical energy and control systems, a more likely consequence than injury is disruption of the operations served by the electrical systems. Whether it is a chemical plant, financial institution, medical facility, mass transportation, or almost any other component of our modern society, an incident resulting in disruption to electrical systems critical to operations can have very significant financial losses. For example, an incident resulting in disruption of electrical energy or control to a hazardous chemical process could result in a process safety event, waste of raw materials, loss of production, and damage to facilities and equipment. A similar incident in a credit card transactions processing center can impact millions of dollars in banking transactions. The hierarchy of hazard control measures in ANSI Z10 *Occupational Safety and Health Management Systems*, coupled with the electrical safety requirements in NFPA 70E, *Standard for Electrical Safety in the Workplace*, serves to prevent incidents that can result in unscheduled disruption of critical electrical systems, thus providing the foundation for operational excellence.

Electrical Mishaps and Collateral Consequences

An incident involving the hazards of slip, trip, fall, confined space, and hazards that have the potential to cause eye, hand, pinching, or crushing injuries can have very serious consequences to the person(s) directly involved, but rarely result in immediate disruption to operations. Electrical incidents differ in a very fundamental way from incidents associated with most other common hazards in the workplace. Incidents involving electrical energy, control, communications, or data systems in general can have far-reaching and immediate consequences, even if there is no injury to the person(s) directly involved. Recognizing this difference can enable an organization to strive for improving the electrical safety program's effectiveness in preventing electrical incidents, even when there may be few or even zero electrical injuries on record.

Whenever there is interaction between people and electrical systems, there is always the possibility for an unexpected or undesirable event to occur. These events have results ranging from serious injury and death to disruptions of some aspect of operations. A misunderstanding or lack of knowledge of how electrical energy can cause injury to the human body is a risk to occupational electrical safety. Without an accurate knowledge of how electrical energy can cause harm, the risk of injury can be underestimated, near-miss incidents may go unreported, and hazard controls undervalued. Any electric shock from a residential, commercial, or industrial power system can cause serious injury, including death. Low-energy shocks can result in internal damage to nerve and muscle tissues, without an observable contact wound. Injuries from low-energy shocks can result in neurological symptoms, nerve pain, and muscle weakness. Arc flash events can result in massive destruction to equipment and extensive injuries from the thermal, acoustic, blast, and electromagnetic energy released. A discussion of electrical injuries is beyond the scope of this book. Recommended references that provide an in-depth discussion of electrical injuries include *Electrical Injuries Engineering, Medical, and Legal Aspects, 2nd ed.*, by Nabours, Fish, and Hill, and *Electrical Safety Handbook, 5th ed.*, by Neitzel, Capelli-Schellpfeffer, and Winfield. The Chicago Electrical Trauma Rehabilitation Institute is one source providing state-of-the-art medical research on electric shock and burn injuries.

Electrical injuries occur at a relatively low frequency, as compared to injuries from other hazards in the workplace. Cawley and Homece noted that electrical hazards are the sixth leading cause of occupational fatalities, but account for only 0.2% of LTIs. This shows that electrical injuries are low frequency but high consequence in nature. The low frequency of LTIs could lead an organization to overlook the need for, and the opportunities in additional benefits of, implementing an effective electrical safety program. Most mishaps in electrical systems do not involve injury, but almost all have undesirable collateral consequences, as shown in Figure 6.1. These disruptions range from annoying with little consequence, to very serious and costly, but with no direct injury.

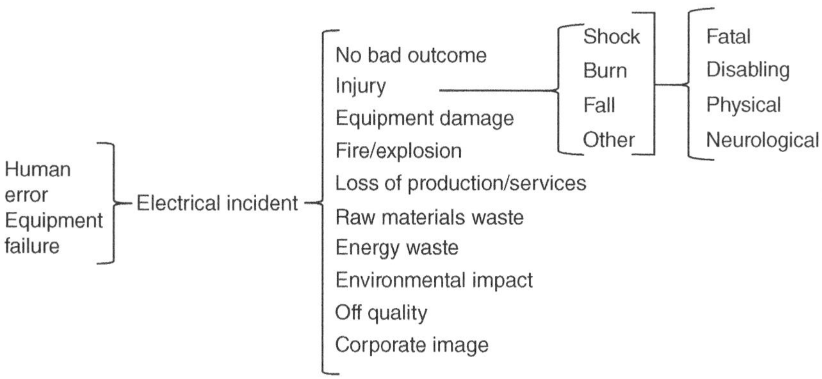

Figure 6.1 Examples of the potential consequences of an incident in electric power or control systems.

Incidents Gleaned From Newspaper Accounts and Incident Reports

Electrical energy, control, communication, and data systems are critical for all aspects of business operations. An incident or mishap in these electrical systems can have significant and perhaps catastrophic impact. The US Environmental Protection Agency reported that in the year 2000, there were 240 chemical releases in the United States due to disruptions in electric power supply. In addition to electrical energy systems, other electrical systems, including instrumentation, monitoring, control, and communication, play critical and vital roles in business operations. The following incidents serve to illustrate that the prevention of electrical safety incidents can have significant impact on operational excellence.

Loss of Normal and Backup Electric Power

During a storm, there was a total loss of electric power supply to the chemical manufacturing plant for 14 minutes, and this caused a fume release through the main stack. Four people out of 26 from the neighboring community that came to the site for evaluation showed slight discomfort as a consequence of this fume release. No employees from the site were affected. The event was covered by the local press. The emergency plan was activated in agreement with established procedures.

The plant electric power was supplied by an on-site cogeneration plant. The site also had a backup supply provided by the local power utility company. Both electrical supplies operated at 13.8 kV, interconnected by means of a tie switch. The configuration of this protection system was designed to provide high reliability of the electrical supply, by isolating either of the two supply systems in the event of a failure or disturbance in it, without affecting the plant operation.

During an intense electric storm, a lightning strike induced a voltage variation in the 115-kV utility supply line 1, to which the plant was connected at the time. The system external utility feeder was automatically transferred to the alternate supply line 2. Simultaneously, the electrical supply from the cogeneration plant was lost due to a shutdown of the two turbines, leaving the site with no power supply.

The 13.8-kV main switch was designed to be remotely opened and closed from the control room of the cogeneration plant. The PLC that controls this operation was powered by a UPS in case of a power failure. In this event, it was not possible to immediately close this switch remotely due to a UPS failure that had not been detected previously.

The plant personnel also went to the main substation and, along with cogeneration personnel, manually closed the 13.8-kV breaker. The plant electric power supply was re-established from the utility company 14 minutes after the original failure.

The cogeneration start-up process needed an external power supply to re-initiate its operation and required approximately 15 minutes to put one of its turbines into operation and begin to supply power to the plant.

The investigation revealed that the instrumentation circuits for the low-low water level interlock in both boilers were not shielded from induced voltages, and a false signal forced the shutdown of the turbo generators.

Missing Grounding and Bonding in a Railcar Unloading Station

An arcing incident occurred at a spent acid railcar unloading station classified as Class I, Division 1, Groups C&D hazardous area. An electrical arc occurred between the railcar and its unloading gangway during a cleaning operation. Fortunately, there were no injuries and no explosion occurred. Following is an excerpt from the incident investigation report:

A spent acid railcar had been unloaded and disconnected. The gangway extension may have had a slight gap between it and the railcar, or it may have been resting on the railcar. A water hose was draped over the gangway handrail and running water to the ground in preparation for rinsing the top of the car. The operator was standing on the gangway extension and heard and saw what looked to be an electrical flash within the neighboring refinery. The lights at the rail spot went out, and the lights in the control room flickered, but no other electrical problems were observed. The operator simultaneously heard and saw an electrical arc out of the corner of his eye between the railcar and the gangway. A few seconds later, another electrical flash was observed at the refinery and the operator heard and saw an electrical arc between the stub at the end of the gangway extension on the right-hand side and the railcar. The lights at the loading rack remained off, and the control room lights flickered again. The operator stepped off the gangway onto the rack. The lights came back on, and the operator finished rinsing the car. The operator did not feel or experience anything out of the ordinary during this incident.

An approximately 400-foot (122-m) longspur rail line was installed for the site. The main rail line is under a utility transmission line for several miles before entering the site. The transmission line has two circuits, a 69-kV circuit about 60 ft (18 m) high and a 13.8-kV circuit about 40 ft (12 m) high. The transmission line is about 40 ft (12 m) horizontally from the unloading station at its closest point and is parallel to the rail line.

The rail line had no bonding jumpers across joints or between the rails, the rails were not bonded to the adjacent loading/unloading structure, and there were no rail isolation joints near the structure. The loading/unloading station structure utilizes foundation earthing and is connected to other earthed structures in the plant.

The refinery powerhouse was contacted, and the incident was discussed. The insulators on a 69-kV transmission line pole located next to the railroad tracks failed. Tracking occurred phase to phase and possibly phase to earth. The first flash was the initial failure, and the second flash was an automatic reclosure of the breaker. The single phase-to-earth fault current is modeled at 18,892 A, the phase-to-phase fault current at 13,750 A, and the three-phase fault current at 15,868 A.

Based on these findings, during the fault at an insulator on the 69-kV line, a portion of the fault current flowed from the site of the fault down to earth, and along the earth back to the 69-kV source. The unloading station's earthing system provided a path for some of this fault return current, and through its loose connection with the gangway and railcar, provided a parallel current path.

Six-Hour Outage in San Francisco Bay Area

A task sequence error resulted in an electric utility crew failing to remove protective safety grounds before re-energizing a section of the utility system that had been de-energized for maintenance. The error resulted in a six-hour power outage in the San Francisco Bay area, impacting millions of customers. Although there was no injury to the workers directly involved in the re-energization tasks, there was a report of an automobile accident death at a street intersection with a non-functioning traffic light.

Millions in Northeast Without Power

An operator error in a power-generating station was attributed as a cause of an extended power outage in August 2003, impacting millions of customers in the eastern United States and Canada.

Delmarva Peninsula Blackout

A wiring error in the modification of a protective control circuit was attributed to causing a power outage lasting for several hours to all customers in the Delmarva Peninsula area on the eastern shore of the Chesapeake Bay.

Hospital Power Outage

One contractor was burned in an arc flash incident, and a hospital was without normal and standby power for four hours. Modifications were being made to the hospital backup power transfer system. A job that was planned to avoid disruption to the hospital electrical systems actually resulted in an extended outage, impacting patients dependent on electrical power for critical care.

Chlorine Release due to Power Outage

During routine functional testing of the utility power-protection systems, an unexpected equipment failure resulted in the complete loss of power to a chemical manufacturing plant. Backup power generators failed to start, and the plant shutdown generated a chlorine release. By chance, wind blew the chlorine cloud offshore, and the surrounding community was not impacted.

In these examples, all had the potential for injury to the workers directly involved. All resulted in highly undesirable and costly collateral impact. Widespread loss of power carries the potential of creating dangerous situations for people far removed from the immediate activity at the root of the incident. For example, the chlorine release could have resulted in danger to the surrounding community.

An organization desiring to improve the performance of electrical systems critical to business operations may find that improving the implementation of an effective electrical safety program could have more benefits than investing in capital projects or additional preventive maintenance. Aeiker et al identified collateral consequences to include process safety, fire safety, and avoidable maintenance and operating costs.

Impact on Operations Requiring High Reliability of Electrical Systems

As illustrated in the incidents earlier, electrical energy, control, communication, and data systems are critical for all aspects of business operations. Reliability of these essential electrical systems is particularly of paramount concern to certain industries, including data centers, hospitals, banking centers, and industries that manufacture, store, transport, and use highly hazardous materials. An incident or mishap in essential electrical systems can have high-cost consequences, even if no injury results. The US Environmental Protection Agency reported that in the year 2000, there were 240 chemical releases in the United States due to disruptions in the electric power supply. In addition to electrical energy systems, other electrical systems, including instrumentation, monitoring, control, and communication, play critical and vital roles in process safety management (PSM).

PSM tends to focus on equipment and systems that are directly associated with hazardous manufacturing processes. Electrical systems, including power, control, and data systems, may not be directly associated with the process; however, the integrity and availability of these electrical systems may be critical and essential for process operations and PSM.

PSM is an essential aspect of managing hazardous manufacturing operations in petrochemical and other industries. Flammable, explosive, and toxic materials used in these manufacturing processes have the potential to cause great harm to workers and surrounding communities. Explosions and fires can cause catastrophic damage to plant and process equipment. A single event can cause irreparable damage to a business. In the United States, regulations are provided by OSHA 1910.119, *Process Safety Management of Highly Hazardous Chemicals, Explosives and Blasting Agents*. In the EU, Council Directive 82/501/EEC, the Seveso Directive provided the initial basis for PSM in Europe. Wellman and McConnell summarized the relationship between PSM and ESM. More recently, the more detailed ATEX Directives 94/9/EC and 1999/92/EC were established.

The task of PSM implementation is complex because it crosses over many functional areas of a business's organization, including R&D, engineering, manufacturing, construction, maintenance, training, and sourcing. An integrated, systematic approach to implementing process safety elements consistently across these business functions is important to manage changes and sustain strong PSM-related performance. Electrical safety in industrial environments is primarily focused on protecting individual workers from harm from electrical hazards, primarily electric shock and arc flash. Like PSM, the task of implementing an electrical safety program is complex because it also crosses over many functional areas of the organization.

Process safety and electrical safety are highly regulated by national governments, and industry standards provide specific guidance for achieving incident-free operations. These two safety competencies are both highly technical in relation to other workplace-related safety issues. Due to the complexity and technical nature, most organizations manage these as distinct and separate. Understanding touch points and overlaps of PSM and electrical safety management (ESM) can result in optimization of safety performance, reliability and uptime of operations, and the use of valuable resources to integrate and optimize PSM and ESM.

There are technology, personnel, and facilities systems that are common to electrical safety, PSM, and other concerns related to business operations.

Technology

Process safety and electrical safety are strongly dependent on applied technology and the fundamentals of chemistry and electricity, and there is an overlap in fundamentals to be considered in a safety program.

Process technology (design basis): Earthing and bonding are essential to both PSM and electrical safety, but perhaps for different reasons. For PSM, earthing and bonding are essential in the control of static energy and sparking across insulating gaps in piping, ducts, and vessels. For electrical safety, the integrity of earthing and bonding is essential to the function of short circuit protective devices and to prevent electric shock from nonconductive metallic structures, equipment, metal cabinets, and enclosures containing electrical components that can become energized in case of a fault. In any case, the installation techniques, maintenance and inspection, and functional testing are the same, whether for PSM or electrical safety purposes.

Hazards analysis: The presence of flammable atmosphere released during normal and abnormal process operation is related to PSM. Electrical installations in those locations to eliminate ignition sources, to specify components and installation techniques, are related to electrical safety. PSM critical equipment control and safe operation configuration, and safety logic diagram are the basis to develop PSM critical systems safety interlocks. Electrical schematic and wiring diagram are key to transforming the design configuration and protection philosophy into reality. This transformation involves close interaction between PSM and ESM.

Procedures and work practices: Operational discipline involves having well-documented procedures, practices consistent with the procedures, implementation without shortcuts, and a process to recognize and manage change outside the boundaries of the established procedures and practices. Procedures should identify overlaps, margins, and touch points of PSM and ESM. For example, a PSM improvement may include modification of electrical hardware and wiring in critical control or monitoring circuits. These changes may also impact electrical safety exposure and risk. This change would need to be evaluated for both PSM and ESM implications.

Management of change – technology: With regard to potentially explosive atmospheres, there must be robust administrative controls to manage changes in process operation, changes of product manufacturers safety data sheets (MSDS), and changes in safety interlocks. Any change in the electrical control systems has potential PSM implications.

Personnel

Personnel training and performance: Electrical safety training is critical to ensuring reliable electrical services for processes. Electrical incidents with fatality due to a shock, injury due to an arc flash, or power outage due to a phase-to-earth fault can shut down a process and initiate events that cause a serious process safety incident.

Contractor safety and performance: Electrical contractors should have demonstrated knowledge, experience, and competency in installing electrical equipment in process safety-managed areas. One example of control required for contractor safety that is common to PSM and electrical safety involves temporary connections to electrical systems, including the temporary connection of voltage testing and diagnostic instruments.

Management of change – personnel: When personnel are changed in the workforce, or when experienced workers retire, key knowledge can be lost. The result can be delays during maintenance, longer shutdowns, and incidents. Whelan et al. described the many issues with the loss of key knowledge in electrical system operation.

Incident investigation and communication: PSM and ESM are distinct and separate, but there is a need to manage overlaps, margins, and touch points. PSM and ESM documents, including procedures and hazards analysis, should be prepared independently by people expert in each competency. In accident or near-miss investigations, PSM experts can miss the electrical contribution to a PSM incident if knowledgeable electrical experts are not participating in the incident investigation. For example, a procedural failure in an interaction with electrical equipment may not be recognized as contributing to a PSM incident.

Emergency planning and response: Electrical systems critical to PSM are not as visible as other major mechanical components, such as large vessels, pumps, reactors, and other process equipment. In many cases, critical electrical equipment is part of the systems of "silent surveillance" that continuously monitors and reacts immediately to emergency conditions. In normal operating conditions, these systems are not needed for production or operation. They may be easily overlooked or forgotten, since it could be years, or even decades, in which they may not operate. Needed for an emergency, the last line of defense can be a system that does not work properly. ESM procedures should address preventive and predictive maintenance of these critical electrical systems. In order for the PSM safeguard system to work, the electrical equipment and systems have to be managed by people competent in electrical technology.

Auditing: Auditing of electrical systems, including personnel, facilities, and technology elements, can find issues that could without resolution become initiators of a process incident. Examples are overheated or overloaded equipment, unlabeled switches, errors in procedures, unsafe work practices, and a lack of management of change for key personnel. Huddleston and Crow discussed the value of auditing in improving the safety and reliability of electrical system.

Facilities

Quality assurance (QA): Electrical equipment must be properly manufactured and installed to work reliably. QA inspections can ensure this. This is critical to PSM and ESM.

Mechanical integrity: Electrical backup power for a PSM critical control system is part of the PSM critical system and is integrated into the mechanical integrity aspects of PSM.

Pre-start-up safety review (PSSR): The use of PSSRs, like audits, can help identify equipment or system problems that may become initiators of a process safety incident, or an electrical incident that could result in multiple consequences, as illustrated in Fig. 1. The electrical system's PSSR provides a final checkpoint for new and modified electrical system/equipment to confirm that all appropriate elements associated with process safety have been addressed satisfactorily and the electrical facility is safe to start up.

Management of subtle changes: Small changes in any system, including electrical, can cause a process problem. Simply changing a fuse to the next higher amperage can change the system's characteristics enough to cause an equipment fault and a shutdown. This can also have a significant impact on the incident energy transfer in an arc flash event in downstream equipment.

The incidents described earlier included either actual or potential safety and/or other operational consequences. Identifying procedures, practices, and administrative controls that serve to prevent common causes provides the opportunity to develop the synergy and optimization of multiple operational objectives, including safety, uptime, operating and maintenance costs, and asset management. Any incident that has an electrical equipment or system failure, or error in the interaction of people with electrical energy, control, communication, or data systems should include an electrical safety resource in the incident investigation. This will help ensure optimum learnings and corrective intervention for all impacted business objectives.

Occupational Safety and Health Management Systems

Table 6.1 provides a partial listing of widely recognized national and international occupational safety management standards.

Table 6.1 Recognized occupational safety management standards.

Standard	Title
ANSI Z10	*Occupational Safety and Health Management Systems, American National Standards Institute*
ILO-OSH	*Guidelines on Occupational Safety and Health Management Systems, International Labour Organization*
ISO 14001	*Environmental management System, International Standards Organization*
OHSAS 18001	*Occupational Safety and Health Management Systems – Requirements, British Standards Institute*
ISO 45001	*Occupational Health and Safety Management Systems – Requirements with Guidance for Use*

These standards are well aligned and provide a framework for managing the policies, planning, implementation, documentation, evaluation, and review of a comprehensive occupational safety and health management system. The managing framework common to these standards includes the "Plan-Do-Check-Act" model to facilitate integration, optimizations, and continuous improvement of a comprehensive occupational safety and health management process. Some companies and organizations have proprietary safety management systems that may be fully aligned with these standards. These proprietary systems may have predated the emergence of national or international standards, or their system may be even more robust than these standards. Whether an organization's safety management system is based on one of the national or international industry standards, or proprietary, it is within this framework that linkages and opportunity for synergies among various safety competencies can be identified and developed.

Synergy and Optimization

The planning and implementation framework of safety management systems is designed to enable integration and optimization of the multitude of safety and health hazards. By establishing an objective to optimize the mutual impact the electrical safety and operational reliability have on each other, an organization can find achievable improvement opportunities. Some examples of areas with mutual impact are described. Other opportunities may be found, dependent on the safety management systems in place in the organization or the type of process or operations.

Contractor Safety Management

When selecting a contractor to perform electrical work in areas where reliability and uptime are of high concern, pre-contract checks must be made to show that the dual competency of electrical engineering and process safety is attained.

Incident Investigations

Incidents having a mishap or failure in electrical systems as a contributing cause must have an electrical safety resource involved in the investigation, even if the potential for an electrical injury is not apparent. Nearly every aspect of operations and an electrical contributing cause could be missed or incorrectly applied.

Design Reviews

Detailed design reviews are done in projects with process safety concerns to ensure that all parts of the process have adequate and essential design. A review of design from the electrical safety viewpoint can ensure that electrical hazards are minimized early in the planning for the project.

Inherently Safer Technology and Designs

The concepts of inherently safer technology and inherently safer processes are generally well established in PSM. The concepts can also be applied to design options in electrical systems.

Process Heat Tracing

If electrical heat tracing is used on a process, an electrical safety review to ensure proper grounding methods can reduce shock hazard exposures. This detailed review may help reduce process incidents from failure of heat tracing energy and control systems.

Grounding, Bonding, and Lightning Protection

Proper grounding and bonding are essential for the operation of overcurrent protective devices and the prevention of electric shock from housings, enclosures, and cabinets of electrical equipment. Grounding and bonding are also essential to the control of static energy in potentially explosive environments. For many PSM-covered processes, due to the presence of flammable liquids, vapors, or dusts, grounding and bonding are key elements of the process design to ensure ignition sources due to static buildup and discharge are eliminated and controlled. Lightning protection systems, important to both personnel safety and process safety, are also dependent on proper earthing and bonding.

Electric Utility and Backup Power Reliability Improvement

Improving the reliability of electrical systems affects electrical safety by reducing worker exposures to electrical hazards during troubleshooting and repair activities. As illustrated in the case history, improving the reliability of electrical systems also reduces exposure to PSM events.

Auditing Processes

Since there are numerous touch points linking electrical safety and process safety, audit processes should be reviewed to eliminate redundancies, overlaps, and voids.

Drawings and Documentation

Revision control of documents is essential to electrical drawings, such as hazardous area classification diagrams, arc flash studies, and single-line diagrams. This method to better control documents for accuracy can be applied to other documentation required for electrical safety, such as work permits, job plans, and audit corrective actions.

Hazard Analysis

As with PSM, it is only by clearly defining and understanding the electrical hazards that we can put appropriate control measures in place to eliminate or reduce the hazard to an acceptable level. Similar methodologies used to conduct process hazards analysis can be used for electrical equipment and facilities.

Emergency Response Planning

Disruption of electrical systems critical to PSM may also create unusual risk for electrical maintenance and operations personnel. Emergency response planning, including emergency drills, should address both electrical safety and process safety consequences of emergency events in electrical systems.

Procedures for Shutdown and Restart

Operating procedures should address normal start-up, normal shutdown, emergency shutdown, and start-up following an emergency shutdown. In many cases, loss of relevant utilities, including electrical power, is considered an emergency. Start-up following such an incident must be done following a thorough review of the safety implications of the shutdown and re-energizing the electrical power.

Crosstalk: Questions for Thought and Discussion

1. What are your management's business objectives?

2. How does the prevention of mishaps in electrical energy, control, and communication systems affect these objectives?

3. Do electrical safety professionals and top management share a common understanding of the total impact of preventing electrical mishaps?

4 How would you describe the value of preventing electrical mishaps?

5 Are you emphasizing the total value when seeking support for your electrical safety program?

Further Reading

Aeiker, J. (2010). Emergency response for electrical safety professionals. *IEEE Industry Applications Magazine* 16 (3): 35–38.

Aeiker, J., Irwin, C., Tomiyoshi, L., et al. (2008). Electrical safety program impact on process safety performance. Conference Record of 5th Petroleum and Chemical Industry Conference Europe: 1–9.

ANSI Z10 (2019). *Occupational Health and Safety Management Systems*. American Society of Safety Professionals.

ATEX Directive 1999/92/EC. European Directive 1999/92/EC (16/12/99) on minimum requirements for improving the safety and health protection of workers potentially at risk from explosive atmospheres.

ATEX Directive 94/9/EC. European Directive 94/9/EC (23/03/94) on the approximation of the laws of the Member States concerning equipment and protective systems intended for use in potentially explosive atmospheres.

Buschart, R.J. (1997). Electrical area classification drawings: a comparison. *IEEE Industry Applications Magazine* 3 (2): 22–25.

Capelli-Schellpfeffer, M., Floyd, H.L., Andrews, J.J. et al. (2004). Safeguarding the electric workplace: an overview of the state-of-the-art in technology, work practices and management systems. *IEEE Industry Applications Magazine* 10 (3): 18–24.

Cawley, J.C. and Homce, G.T. (2003). Occupational electrical injuries in the United States, 1992–1998, and recommendations for safety research. *Journal of Safety Research* 34 (3): 241–248.

Doan, D.R., Fannon, J.S., Floyd, H.L., and Scarborough, M. (2009). Railroad tracks: a potential source of ignition in hazardous (classified) locations. *IEEE Industry Applications Magazine* 13 (2): 62–67.

Cadick, J., Capelli-Schellpfeffer, M., Neitzel, D.K., and Winfield, A. (2019). *Electrical Safety Handbook*, 5e. New York: McGraw Hill.

Floyd, H.L. (1986). Reducing human errors in industrial electric power systems operation – part 1: improving system reliability. *IEEE Transactions on Industry Applications* 1A.- 22 (3): 420–424.

Hill, P.F., Fish, R., and Nabours, R.E. (2004). *Electrical Injuries Engineering, Medical, and Legal Aspects*, 2e. Tucson, AZ: Lawyers and Judges Publishing Co.

Huddleston, R.L. and Crow, D.R. (2005). A second set of eyes. *IEEE Industry Applications Magazine* 11 (3): 63–68.

Kolak, J. (2007). Electrical safety – elements of an effective program. *Professional Safety* 52 (2): 18–24.

Leroux, P. (2007). New regulations and rules for ATEX directives. *IEEE Industry Applications Magazine* 13 (1): 43–51.

Liggett, D. (2007). Training and qualifying your employees. *IEEE Industry Applications Magazine* 13 (3): 25–30.

McCollum, K., Gowrishankar, T.R., and Lee, R.C. (2022). Hidden realities of electrical injuries. In: *Nursing*. Wolters, Kluwer Health Inc.

Neitzel, D.K. (2007). Electrical hazard analysis. *IEEE Industry Applications Magazine* 13 (3): 12–16.

NFPA 70E. *Standard for Electrical Safety in the Workplace*. National Fire Protection Association.

US Environmental Protection Agency. Chemical accidents from electric power outages, Bulletin EPA 550-F-01-010, September 2001, 2024.

Wellman, C.M. and McConnell, J.D. (1997). Process safety management and the electrical function. *Conference record of the 1997 IEEE IAS Petroleum and Chemical Industry Technical Conference* (September 1997), pp. 109–119.

Whelan, C.D., Aeiker, J., Downey, S., and Long, K.L. (2004). A looming crisis? [Industrial electric power skills competency]. *IEEE Industry Applications Magazine* 10 (2): 51–56.

7

The Value of Vulnerability

Introduction

> How could this have happened? We have the best safety record in our division. We've never had a lost time electrical injury at this plant. He is the most knowledgeable person in the crew. He has never had a recordable injury. I was not aware that my employees were exposed to a hazard with such severity. I don't understand.
>
> –A plant manager commenting during an arc flash injury investigation

> I've been in the business (electrical trade) for 25 years. Until today, I can honestly say that I've never understood the hazards I've worked with.
>
> –An electrician at the end of an eight-hour hazard awareness seminar

These two quotes may sound familiar to electrical safety professionals. The first is from a plant manager during the course of an investigation of an arc flash incident that resulted in an LTI and long-term disability to a worker. In the statement, we see how the plant manager's disbelief is entangled in his justification of the worker's "qualifications" as the "most knowledgeable person in his crew." In short, the manager does not understand how such a horrible event happened to a person whose knowledge of rules and regulations was so complete.

As if in response to the questions posed by that manager, the second quote is from an electrician who had worked in the field for more than 25 years and was well-versed in electrical safety regulations and various employers' safety rules (which was what had comprised most of his safety training thus far). His co-workers and employers had considered him to be a "qualified" electrician. This electrician admitted that he had worked his entire career up to that point without fully understanding the hazards of electricity. Through a course on electrical hazards, he had come to realize that his evasion of electrical injury had been due not so much to following the standards of the field but to luck.

Hidden Risk in Occupational Electrical Safety: Through the Lens of System Safety,
First Edition. H. Landis "Lanny" Floyd.

Perceived Vulnerability

How do we best improve the quality of worker training to result in workers' understanding of their personal vulnerability? In other words, to help workers identify with the electrician quoted from the end of the hazard awareness seminar. This level of awareness ultimately impacts their tactful adoption of safety protocol. The quality of training needs assessments, content design, and delivery methods are important factors for training effectiveness.

Safety professionals have already contributed greatly to the understanding of good training practices. These topics are the featured sessions at safety conferences and articles in professional literature, which provide foci ranging from the importance of understanding how adults learn as well as understanding how workers cognitively process risk perceptions. For example, Fanning provides an excellent and practical discussion of techniques for engaging adult learners (e.g., role-playing, group projects, guided learning, storytelling, and peer coaching), because adults learn differently than children and, therefore, do not respond well to content-focused education. Merli summarized key considerations in designing content and delivery for adult learners and suggested four different levels of training evaluations (learner's reaction, learning, behavior change, and return on investment (ROI)/decreased incidents/increased profits). Cullen helps us recognize the great importance of truly understanding employees' workplace cultures before developing training programs and approaching safety training in high-risk environments through personal stories. Lehmann, Haight, and Michael reported on the effects of safety training on risk tolerance and concluded that workplace safety training alone should not be considered adequate enough to expect appropriate risk decision-making to result among employees. They suggest that "more specialized (i.e., psychological or behavior-based) training is necessary for changing safety-related attitudes and behaviors."

To use a grave illustration of why it is important to highlight the risk perceptions of workers, consider the subset of workplace hazards that constitute a relatively small percentage of nonfatal injuries but have a significantly higher likelihood of causing disabling or fatal injuries (serious injuries and fatalities [SIF] potential). The injuries they produce are commonly referred to as low-frequency/high-consequence injuries. (For example, compare rates of *Fires and Explosions* and *Contact with Electricity* from Tables 7.1 with 7.2.) Although exposure to the hazards may be routine, the low frequency of nonfatal injury (Table 7.1) can create an illusion of control or immunity, an illusion shattered by the very serious consequences of these hazards, as illustrated in Table 7.2, and brought to life by the two quotes from our Introduction. To that end, a safety training and assessment that focuses only on risk likelihood without a focus on

Table 7.1 Comparing select nonfatal lost-time injuries (US Bureau of Labor Statistics data.)

Type of Nonfatal Injury	Number of Injuries (2010)
Total	1, 191, 100
Sprains, strains, tears	474, 000
Musculoskeletal disorders	346, 300
Falls on the same level	182, 400
Struck by object	138, 530
Falls to lower level	73, 520
Assault/violent act by person	40, 310
Highway accidents	36, 460
Assault/violent act by animal	7, 160
Fires and explosions	3, 000
Electrical shock and burn	1, 890

Table 7.2 Ratio of lost-time injuries to fatal injuries.

Event or Exposure	LTI/Fatality Ratio[a)]
Fires and explosions	12
Contact with electricity	13
Transportation accidents	23
Assaults and violent acts	28
Fall to lower level	104
Caught in, compressed or crushed	134
Struck by object	323
Falls on the same level	2, 056
Struck against object	8, 414
Slips or trips without fall	12, 593
Overexertion in lifting	14, 033

a) Lost-time injury to fatality ratio.
Sources: Anderson and Denkl and Cawley and Brenner.

risk severity is a disservice to our workers. Safety training and assessment should focus on both risk likelihood *and* severity.

Where Does Electrical Safety Begin?

> Why should my company invest in developing safer electrical equipment, when you can eliminate the hazard by creating an electrical safe work condition?
>
> –A senior vice president of an electrical equipment manufacturer

Unfortunately, the executive who made this comment did not understand the facilities life cycle, shown in Figure 7.1, and the powerful impact engineering design solutions have on reducing the risk of occupational injuries. His perception of risk and the effectiveness of measures to control risk was flawed. The need to identify and control risks does not start with a worker confronting an electrical hazard in the workplace. Electrical safety begins in the conceptual design phase of tools, equipment, systems, and facilities. For more than 70 years, safety risk management literature has included the concept of applying a hierarchy of risk controls that begin in the conceptual design and extend through the operational phases of tools, equipment, systems, and facilities.

Choices to utilize higher order controls in the design phase reduce the dependence on the lower order controls (warnings, administrative controls, and PPE). The effectiveness of lower order controls relies on the real-time knowledge and action of employees in managerial and supervisory roles, as well as real-time knowledge of the workers at risk themselves. The relatively lower effectiveness of these controls is a function of the variability in people's ability to pay attention and respond appropriately when using warnings, administrative controls, and PPE.

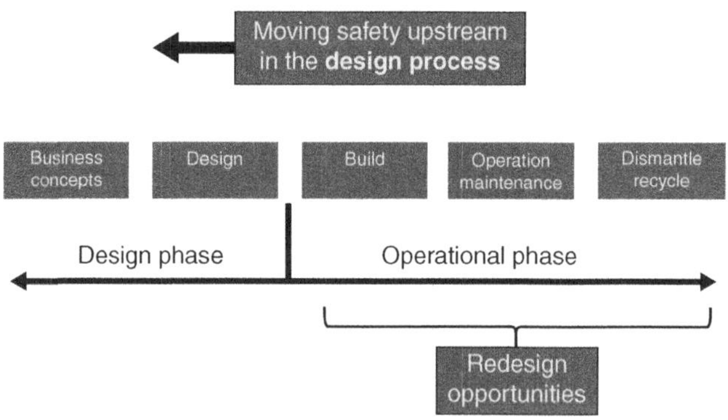

Figure 7.1 The facilities life cycle.

Selection of these lower order controls is typically made during risk assessments after design is complete to address residual risks not reduced ALARP in the design process. While lower order controls depend on workers' attention to recognizing hazards and doing the right thing at the right time to avoid injury, research from cognitive psychology (the field of psychology that studies attention) demonstrates that it is simply not possible for a human being to focus attention at all times and sustain attention for long periods of time. These limitations of attention are not flaws in human behavior; it is a natural process of being human. The nature of human attention highlights some explanations of residual risk present in lower order controls and, therefore, offers support for the importance of well-developed higher order controls in potentially hazardous work environments.

Effective application of risk controls during the design phase requires engineers, designers, and project managers (who are removed from the direct line of fire for hazards in the workplace) to understand worker vulnerability and the limits of human performance, including attention.

Likelihood

Injuries from electric shock and arc flash are relatively rare in the workplace, which can lure us into the trap of focusing only on risk likelihood but not on both likelihood and severity. Data from the United States indicates that, in recent years, nonfatal electrical injuries account for less than 0.2% of all nonfatal LTIs. The very low frequency of electrical injuries relative to the total of all occupational injuries may lead an individual or organization to conclude that the low likelihood of injury from an individual task or exposure may be acceptable. However, a culture that accepts such low-likelihood short-term risks may not fully understand the long-term risk. To illustrate this point, consider a scenario in which the likelihood of a severe outcome when performing a particular task is 1 in 1,000. An individual or organization may consider that to be an acceptable risk. But what if that individual or organization made a similar decision every day for 20 years? What is the likelihood that the individual could suffer a life-changing injury in their 20-year working career? The likelihood can be calculated as follows: The number of events, n, is (20 years)(5 days/work week)(40 work weeks/year) = 4,000. The assumed likelihood (or probability) for each independent event is 0.001.

The probability of a severe outcome during the 20 years:

$$1 - (1 - \text{probability})^n$$

$$1 - (1 - 0.001)^{4000} = 1 - 0.01827 = 0.98173$$

Table 7.3 Three dimensions of risk perceptions. The examples offered in the last column depict possible thoughts of a person who does *not* perceive themselves to be at risk.

Dimension of Risk	Description of Dimension	Example Thought from a Hypothetical Individual
Perceived likelihood	Individual's perception of the probability a hazard may occur.	I have a low likelihood of encountering an incident on the job today.
Perceived susceptibility	Individual's perception of vulnerability to a hazard.	I am not likely to get burned on the job, even if an indecent occurred.
Perceived severity	Individual's perception of the degree of harm that a hazard would cause.	If I were to be burned, it wouldn't be so bad.

Adapted from Brewer et al. [1].

In this hypothetical scenario, there is a 98% probability that by accepting one low likelihood risk every day, the worker will suffer a serious injury at some point during their 20-year working career.

An understanding of what is meant by "risk perception," a psychological construct, is foundational to the ideas we are sharing in this chapter. To fully understand "risk perception," we need to look at several dimensions of risk. Clemens and Simmons propose a focus on the concepts of risk severity and probability (probability being similar to likelihood). Lehmann et al. propose a focus on the concepts of risk tolerance and risk perception. Understanding risk must come out of three dimensions of personal perception: perceived likelihood, perceived susceptibility, and perceived severity of the hazard (see Table 7.3). Training that provides information about rates of incident, or degree of injury, but not on a personal connection to the numbers or images used in the training, may miss some of these dimensions of risk.

Consider how someone developing a training module on construction jobsite electrical safety might integrate the three dimensions of risk in a hypothetical training quiz question.

What is the reason GFCIs are required when using any electrically powered hand tool on the jobsite?

A) OSHA 1926 requires the employer to provide GFCI-protected outlets on construction sites.
B) The National Electrical Code requires GFCI-protected outlets to be installed on construction sites.
C) NFPA 70E Standard for Electrical Safety in the Workplace requires that all extension cords and portable electrically powered tools be plugged into outlets having GFCI protection.
D) A GFCI will protect a worker, including myself, from instant death by electrocution due to a defective tool or cord.

There are no "wrong" answers in this case, as each of choices A through D is technically correct. However, the first three choices are about understanding rules and regulations. The last one, choice D, is an indicator of understanding risk and personal vulnerability.

Here is a second example, representing training on auditing the use of GFCIs on a construction site.

Which of the following items provides better assurance that workers are complying with requirements to use GFCIs?

A) When observing workers using portable electric-powered tools, determine if GFCIs are being used.
B) When observing workers using portable electrical tools, engage the workers and determine if they understand the purpose of GFCIs.

Again, while there are no "wrong" answer choices, a positive response on B is likely a better indicator of both the workers' commitment to always use a GFCI and the auditor's understanding that workers' behaviors when being observed are not necessarily representative of their normal behavior.

What answer choice D in the first example and answer choice B in the second example have in common is that they have better made a personal connection toward why the rule is in place; a personal connection that builds on risk likelihood, susceptibility, and severity.

Now that we have addressed the construct of "risk perceptions" to include three dimensions of risk, the next section introduces some theoretical models from the field of psychology, explaining why establishing a personal sense of vulnerability (i.e., integrating all three dimensions of risk) is an essential precursor to workers' ultimate goal of understanding and committing to follow safety rules and procedures.

Of course, our main goal in providing safety training is to influence behavior. We want people to behave safely on the job. Behavior change is well studied in health psychology, a field that has studied behaviors from smoking cessation to environmentally sustainable actions.

A Look at Theory

Two of the classic health psychology models that outline the relationship between risk perception and behavior are the Health Belief Model (HBM) (described thoroughly by Becher and Maiman [2]) and the Theory of Planned Behavior (TPB), first developed by Ajzen [3].

The HBM (see Figure 7.2) helps us understand a person's *perception of risk related* to a hazard and *perception of benefit* related to a protective behavior. These factors together influence whether a person is likely to use a protective behavior to prevent injury.

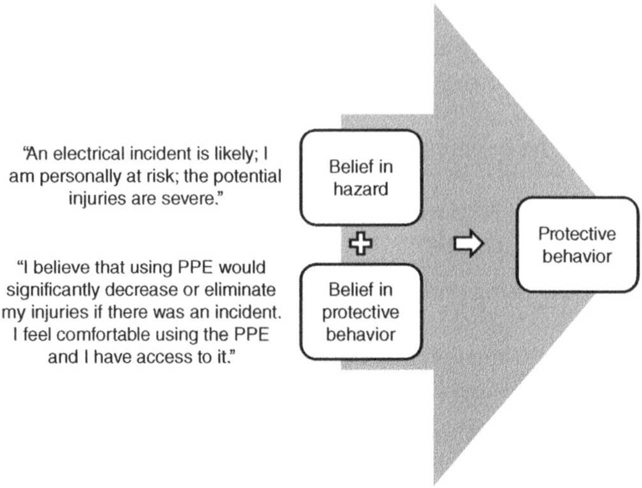

Figure 7.2 Health Belief Model. The Health Belief Model (simplified and adapted from Becher and Maiman) is, in a sense, a psychological cost–benefit analysis of a potential threat and a possible behavioral solution to that threat. You can see the three dimensions of risk from Table 7.3 (likelihood, susceptibility, and severity) captured in the Belief in Hazard portion of the model. The example thoughts offered on the left side depict a person who *does* perceive themselves to be at risk and *does* perceive that protective behavior will be a benefit. Figure created by Floyd.

The TPB (see Figure 7.3) describes the psychological, social, and environmental factors that influence whether a person intends to change their behavior. Importantly, it outlines the factors that affect one's *intention* of changing behavior (i.e., intention precedes action). According to the TPB, there are three main factors that help us understand a person's intention to adopt a safety behavior, such as using PPE: (1) their belief that the protective behavior will benefit them; (2) whether their social circle supports the protective behavior; and (3) whether they feel confident knowing how to carry out the protective behavior.

Both of these theories illustrate clearly the need for a person to perceive the following before they intend to adjust their behavior:

1) Their risk (likelihood/susceptibility/severity) is high.
2) Self-protective behavior (e.g., PPE) will lessen the likelihood/susceptibility/severity.

This is valuable information that we can incorporate into our safety curriculum and leads to the question: How can we develop, in our audience, a sense of vulnerability during a training? The answer to this question comes from an understanding of how risk perceptions are formed.

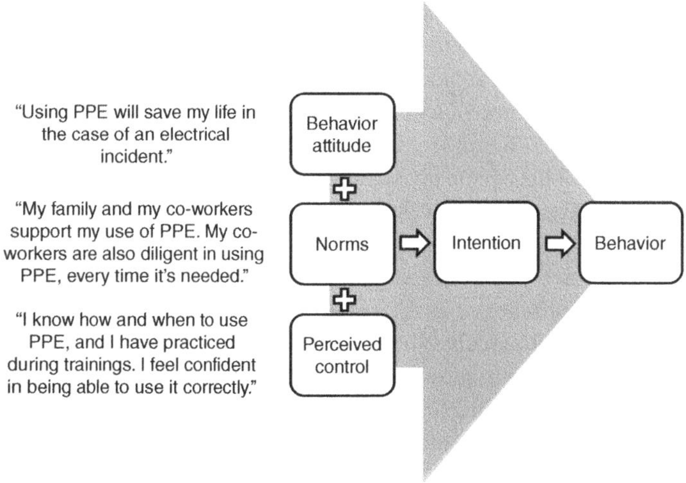

Figure 7.3 Theory of Planned Behavior. The Theory of Planned Behavior (here, adopted from Ajzen) helps us understand a person's *intention to adopt a protective behavior*. The example thoughts offered on the left side depict possible thoughts of a person who does perceive themselves to be at risk, feels supported and confident in their use of protective behavior, and is thus likely to intend to use protective behaviors. Figure created by Floyd.

Psychological models of behavior change tell us that intention to behave safely will only occur if a person believes there is a high-risk likelihood, susceptibility, or severity and also believes that self-protective behavior (e.g., PPE) will lessen the likelihood, susceptibility, or severity of the hazard. In this section, we share some background on how people form risk perceptions. This will be valuable for developing techniques you can use in training to help workers appreciate the risk posed by electrical work and the protective power of safety practices.

Risk Perceptions Are Primarily *Feeling Based*

When we make a judgment about risk likelihood and risk severity for events, including an adverse event like a workplace injury, there are two mental systems we use. One is an analytic system (which is logic oriented, governed by conscious thought processes, and time-consuming). The other is an experiential system (which is feeling-based, governed by associations and "vibes" from previous experiences, and quick).

Because it is efficient (though not necessarily accurate), human beings rely *primarily* on the experiential, feeling-based system to make decisions. We are particularly likely to rely on the experiential system when decision-making is complicated, when we have little time to make a decision, and when our mental resources are taxed through fatigue, hunger, or other types of preoccupation.

The experiential system that we typically use to calculate risk relies heavily on past experience. When we have a strong emotional experience associated with certain hazards, this has a profound impact on our risk perception. For example, a person who has been on the scene of a fatality that was due to another worker's noncompliance with proper PPE protocol will associate noncompliance with feelings connected to trauma. This person does not need much more convincing to be persuaded to be vigilant regarding safety. A person who has never been present at a serious injury or fatality will need more convincing.

Because adoption of safety behaviors is strongest when a person perceives a high-risk likelihood, susceptibility, or severity along with a belief that safety behaviors will lessen that likelihood, susceptibility, or severity of a hazard, it follows that we should address this important question: How can we establish, in our safety training, a perception of high-risk likelihood, susceptibility, and severity (i.e., vulnerability) along with the belief that safety behaviors will help?

We operate on the experiential (feelings-based) system when developing our own risk perceptions, and when making our own risk decisions, it makes sense to use methods that draw on feelings to increase our workers' perceptions of risk prior to a safety training. This can occur alongside more traditional methods (e.g., presenting statistics).

The Role of Stories

The experiences and opinions of other people play a key role in how we perceive appropriate actions and behaviors. We can see this clearly in the models presented above, specifically the subjective norms factor in the TPB. Our value of others' experiences and opinions is also very well understood through social learning theory, a construct developed in the 1950s by social psychologist Albert Bandura. Bandura's theory outlines how people learn vicariously through watching others, a characteristic foundational to human learning and most obvious during childhood when human children are hard-wired to mimic adults' behaviors. Social learning is, of course, not limited to childhood and continues throughout our lifetime.

Social learning theory explains why we learn vicariously through watching others, but it can also be applied to learning through others via *narratives*, or stories. Health and safety messages presented in the form of stories have been studied extensively to determine the effect they have on subsequent behaviors. Stories can come in different forms, ranging from journalism to the literature to testimonials, and can depict occurrences to characters that are either fictional or real.

Of great importance to our purposes of health and safety communication, *the more similar people perceive themselves to be to a narrative character, the more likely*

A. "There are 1,890 electrical shock and burn injuries each year."

B. "Last Friday, one of our linemen was critically injured in an electric shock incident. His spouse and their two children have spent the week visiting him in the hospital."

Figure 7.4 Narrative, stories, and perceptions of risk. These are two hypothetical examples that could be used in a company newsletter. Which of these two examples is more likely to persuade your workers to take appropriate safety measures? Item B mentions an individual's spouse and two children – most workers will be able to project themselves into that type of story, making it more likely to lead to behavior change in comparison to the presentation of statistics only. Note that the information given is not *overly* detailed, which would make the story more difficult for another person to identify with.

they are to be persuaded by the narrative itself. The results are equivocal in whether statistical or narrative types of communication are more persuasive. With this in mind, it may be best to offer both strategies in a communication message.

Presenting Narratives

Stories of people affected by incidents that include names and references to personal characteristics (e.g., family and hobbies) will be more persuasive than presenting only statistics. The more a person can see themselves in the shoes of another, the more likely they are to be transported into and affected by the story; that is, the more likely they are to think, "that could be me." Creating these stories is an art; if too much detail is included, people can no longer project themselves in the story.

When possible (and with permission from those directly impacted), safety training materials can use real-life stories to establish a personal sense of vulnerability that is needed to achieve behavior change. In these stories, personable information that is generic enough to apply to many people (see Figure 7.4) can be valuable.

Because safety-related training is an essential component of hazard and risk management, but can entail significant costs and takes people out of their daily jobs, it is important to make the best possible use of that time. Based on theoretical models of behavior change used in psychology, establishing a personal sense of vulnerability during safety training can help employees adopt self-protective behavior. Training curricula should include stories (in addition to statistics) to establish a feeling of vulnerability because people develop risk perceptions through experiential (feeling-based) processes.

Crosstalk: Questions for Discussion

1. How does your organization educate workers on the hazards of electricity, and how an electrical injury would impact their lives?
2. What routine tasks in your operations have potential exposure to electricity greater than 50 V?
3. How are managers and supervisors educated on potential exposures that occur routinely in normal operations?
4. What could you do to increase reporting of near-miss electrical incidents?
5. How could your organization utilize materials available from the Electrical Safety Foundation to increase awareness and understanding of the unrecognized risk of electrical injuries?

References

1 Brewer, N.T., Chapman, G.B., Gibbons, F.X. et al. (2007). Meta-analysis of the relationship between risk perception and health behavior: The example of vaccination. *Health Psychology* 26 (2): 136–145.
2 Becher, M.H. and Maiman, L.A. (1975). Sociobehavioral determinants of compliance with health and medical care recommendations. *Medical Care* 13 (1): 10–24.
3 Ajzen, I. (1985). From intentions to actions: A theory of planned behavior. In: *Action-control: From cognition to behavior* (ed. J. Kuhl and J. Beckman), 11–39. Heidelberg: Springer.

Further Reading

Floyd, A.H.L. and Floyd, H.L. (2014). The Value of Vulnerability: Helping Workers Perceive Personal Risk. *Professional Safety Journal of the American Society of Safety Professionals* 59 (4): 32–37.

Fanning, F.E. (2011). Engaging learner techniques to make training stick. *Professional Safety* 56 (8): 42–48.

Merli, C.M. (2011). Effective training for adult learners. *Professional Safety* 56 (7): 49–51.

Cullen, E.T. (2011). Effective training: A case study from the oil & gas industry. *Professional Safety* 40–47.

Lehmann, C.C., Haight, J.M., and Michael, J.H. (Spring 2009). Effects of safety training on risk tolerance: An examination of male workers in the surface mining industry. *Journal of SH&E Research* 6 (1).

Floyd, A.H.L. and Floyd, H.L. (2014). Cultural drift and the occlusion of electrical safety. *IEEE Trans. Ind. Appl.* 50 (3): 1610–1618.

Floyd, H.L. (2015). A Practical Guide for Applying the Hierarchy of Controls to Electrical Hazards. *IEEE Transactions on Industrial Applications* 51 (4): 4263–4264.

ISO 45001. *Occupational Health and Safety Management Systems.* Geneva, Switzerland: International Standards Organization.

Floyd, H.L. and Floyd, A.H.L. (2017). Residual Risk and the Psychology of Lower Order Controls. *IEEE Transactions on Industry Applications* 53 (6): 6009–6014.

Floyd, H. L, When maintenance error is the cause of electrical injury – Lessons from aviation safety, *2019 IEEE IAS Electrical Safety Workshop*, Jacksonville, Florida, March 4–8, 2019.

Clemens, P.L. and Simmons, R.J. (1998). *Safety System and Risk Management, A Guide for Engineering Educators.* U. S. Department of Health and Human Services (NIOSH) Publication No. 96-37768.

Ajzen, I. (1991). The Theory of Planned Behavior. *Organizational Behavior and Human Decision Processes* 50: 179–211.

Ajzen, I. and Fishbein, M. (1980). *Understanding Attitudes and Predicting Social Behavior*. New Jersey: Prentice-Hall.

Slovic, P., Finucane, M.L., Peters, E., and MacGregor, D.G. (2004). Risk as analysis and risk as feelings: Some thoughts about affect, reason, risk, and rationality. *Risk Analysis* 24 (2): 311–322.

Watson, D. & Clark, L.A. (1994). THE PANAS-X: Manual for the positive and negative affect schedule – expanded form. Retrieved online August 10, 2012 from www.psychology.uiowa.edu/faculty/clark.

Weinstein, N.D. (1989). Effects of personal experience on self-protective behavior. *Psychological Bulletin* 105: 31–50.

Finucane, M.L., Alhakami, A., Slovic, P., and Johnson, S.M. (2000). The affect heuristic in judgments of risks and benefits. *Journal of Behavioral Decision Making* 13 (1): 1–17.

Bandura, A. (1977). *Social Learning Theory.* Oxford, England: Prentice-Hall.

Kreuter, M.W., Green, M.C., Cappella, J.N. et al. (2007). Narrative communication in cancer prevention and control: A framework to guide research and application. *Annals of Behavioral Medicine* 33 (3): 221–235.

Hinyard, L.J. and Kreuter, M.W. (2007). Using narrative communication as a tool for health behavior change: A conceptual, theoretical, and empirical overview. *Health Education & Behavior* 34 (5): 777–792.

Weinstein, N.D., Marcus, S.E., and Moser, R.P. (2005). Smokers' unrealistic optimism about their risk. *Tobacco Control* 14: 55–59. https://doi.org/10.1136/tc.2004.008375.

Geer, L.A., Curbow, B.A., Anna, D.H. et al. (2006). Development of a questionnaire to assess worker knowledge, attitudes and perceptions underlying dermal exposure. *Scand J Work Environ Health* 32 (3): 209–218.

Rundmo, T. (1996). Associations between risk perception and safety. *Safety Science* 24 (3): 1996.

8

Safety Management Systems

Safety is not the absence of incidents and injuries, but about the capacity to ensure continuous improvement in processes to reduce the risk of injury to ALARP.

OSHA electrical safety-related regulations provide requirements for safe work practices, and NFPA 70E *Standard for Electrical Safety in the Workplace* reinforces the OSHA requirements with detailed guidance on implementing safe work practices. The requirements and guidance in these resources are essential to your electrical safety program but do not provide a comprehensive solution necessary to effectively manage electrical safety in the workplace. If this limitation is not understood, the efforts to ensure compliance with regulations and industry electrical safety standards will likely not be as effective in preventing serious injuries and fatalities as if compliance is achieved within the context of a state-of-the-art safety management system.

Safety management systems help ensure a comprehensive approach to managing exposure to hazards in the workplace. Safe work practices, which are the primary content of OSHA regulations and NFPA 70E, are just one part of a comprehensive electrical safety program. Recent revisions to NFPA 70E added references to ANSI/ASSP Z10, *Occupational Health and Safety Management Systems*, and ISO 45001, *Occupational Health and Safety Management*, noting the value of safety management system standards to help ensure a comprehensive and effective electrical safety program.

Excellence in electrical safety, as with all business objectives, requires thoughtful planning and shared commitment throughout the organization. In this context, planning is not about the essential planning as in task and job planning. Planning for an effective safety management system involves deciding how the organization will identify and prioritize electrical safety concerns, establishing goals and objectives to guide improvement, and creating implementation plans, including ensuring financial and personnel resources are sufficiently allocated to achieve

Hidden Risk in Occupational Electrical Safety: Through the Lens of System Safety,
First Edition. H. Landis "Lanny" Floyd.

success. The planning process outlined in ANSI/ASSP Z10 consists of four main steps:

1) Identifying OHS issues within the organization, including hazards, risks, known deficiencies, and improvement opportunities.
2) Prioritizing the identified issues based on their importance to the employer.
3) Establishing objectives to improve the effectiveness of safety initiatives intended to reduce risk.
4) Creating a safety management implementation plan that specifies tasks and assigns responsibilities to meet objectives.

ANSI/ASSP Z10 and ISO 45001 identify core elements of the process to help ensure effective planning, implementation, and continuous improvement.

Core Elements of OSHMS Standards

The core elements of OSHMS standards are interrelated and are applicable to all occupational safety and health risks. The application of the core elements is intended to help ensure a consistent approach to plan, design, control, and monitor efforts to continuously reduce risk. Seven core elements are discussed later to illustrate a comprehensive electrical safety program that includes safe work practices in OSHA regulations and NFPA 70E and safety management requirements of OSHMS standards.

Management Leadership

Management provides the leadership, vision, and resources needed to implement an effective electrical safety program. Top management should demonstrate its commitment to continuous improvement in electrical safety, communicate that commitment to workers, and set program expectations and responsibilities. Managers at all levels should make electrical safety a core organizational value, establish safety and health goals and objectives, provide adequate resources and support for the program, and set a good example by demonstrating a commitment to eliminating hazards, reducing risks, protecting workers, and continuously improving workplace electrical safety. Top management must demonstrate its commitment to continuous improvement in electrical safety, communicate that commitment to workers, and set program expectations and responsibilities.

Some examples of visible leadership include establishing a written policy describing the organization's commitment to electrical safety; communicating the policy to all workers and stakeholders, including contractors, suppliers, and visitors; providing the resources needed to implement the electrical safety program; integrating electrical safety into planning and budgeting processes; and aligning budgets with program needs.

Worker Participation

Workers should be involved in all aspects of the program – including setting goals, identifying and reporting hazards, investigating incidents, and tracking progress. All workers, including contractors and temporary workers, understand their roles and responsibilities under the program and what they need to do to effectively carry them out. By encouraging workers to participate in the program, management signals that it values their input into electrical safety decisions. Workers are often best positioned to identify electrical safety concerns and program shortcomings, such as emerging workplace hazards, unsafe conditions, close calls/near-misses, and actual incidents. By encouraging reporting and following up promptly on all reports, employers can address issues before someone gets hurt. Including worker input at every step of program design and implementation improves the ability to identify the presence and causes of workplace hazards, creates a sense of program ownership among workers, enhances their understanding of how the program works, and helps sustain the program over time. Opportunities to engage workers span all aspects of the program, including setting goals, identifying and reporting hazards, identifying safe work practices, developing safety procedures, participating in incident investigations, training electrical and nonelectrical workers, and leading safety audits and program evaluation. Workers must feel that their input is welcome and that their voices will be heard. Differences in language, education, or skill levels in the workplace must be considered.

Hazard Identification and Assessment

- Procedures are put in place to continually identify workplace electrical hazards and evaluate risks.
- Electrical safety hazards from routine, nonroutine, and emergency situations are identified and assessed.
- An initial assessment of existing hazards, exposures, and control measures is followed by periodic inspections and reassessments to identify new hazards.
- All incidents, including near-misses, are investigated with the goal of identifying the root causes.
- Identified hazards are prioritized for control.

A common root cause of workplace electrical injuries is the failure to identify or recognize hazards that are present or that could have been anticipated. Unrecognized hazards or underestimation of risk can lead to unintentional normalization and acceptance of risk. A critical element of any effective electrical safety program is a proactive, ongoing process to identify and assess such hazards. Hazards can be introduced over time as work areas and processes change, equipment or tools are damaged, cords become worn, maintenance is neglected, or housekeeping practices decline. The best time to investigate an electrical incident is *before* it occurs. Setting aside time to inspect the workplace for electrical hazards regularly can help identify shortcomings so that they can be addressed before an incident occurs. For each hazard exposure identified, you should determine the severity and likelihood of incidents that could result and use this information to prioritize corrective actions. Some hazards, such as damaged cords or missing covers, should be fixed when they are found. Fixing hazards on the spot emphasizes the importance of electrical safety and takes advantage of a safety leadership opportunity.

You should include all areas and activities in these inspections, such as storage and warehousing, facility and equipment maintenance, purchasing and office functions, and the activities of onsite contractors, subcontractors, temporary employees, and visitors. Checklists that highlight things to look for are helpful but should not be allowed to limit inspections.

Checklists can be developed for major categories, such as work environment, equipment operation, equipment maintenance, work practices, and work organization.

Workplace electrical incidents, including injuries, close calls/near-misses, and equipment failures, can provide a clear indication of where hazards exist. By thoroughly investigating incidents and reports, hazards that are likely to cause future harm are identified. The purpose of an investigation must always be to identify all possible contributing causes to identify every opportunity to prevent future occurrences. Having a plan and procedure for conducting incident investigations can enable investigations to begin immediately when an incident occurs. The investigation plan should cover who should be involved, what skills are needed to be available, and what training members of the investigation team need.

Risk Control

When hazards are identified, controls should be selected according to a hierarchy that uses engineering solutions first, followed by safe work practices and other administrative controls, and, finally, PPE. Employers should select the controls that are the most feasible, effective, and permanent. New technologies may have the potential to be more protective, more reliable, or less costly. Whenever possible, select equipment, machinery, and materials that are inherently safer based on PtD principles. Apply PtD when making your facility, equipment, or product design decisions.

It is crucial to ensure that selected controls are implemented, interim protection is provided, progress is tracked, and the effectiveness of controls is verified. Evaluate control measures to determine if they are effective or need to be modified.

Involve workers in the evaluation of the controls. If controls are not adequate, identify, select, and implement further control measures that provide adequate protection. For example, after establishing an electrically safe work condition, hazardous energy may still exist at the boundary of the electrically safe working condition, such as on the line side of isolation devices, in adjacent compartments of motor control centers or switchgear, behind shutters in draw-out circuit breaker cells, and behind doors and covers that could be opened. Risk control planning should take into consideration how miscommunications, distractions, or misunderstandings could lead to failure of the implemented controls to prevent exposure or injury. By asking, "How could this control fail?," additional engineering or administrative controls may be warranted to control the remaining risk. It is usually the case that no single method fully protects workers; requiring a combination of controls or multiple layers of protection is almost always appropriate.

Education and Training

Education and training are essential tools for informing workers, supervisors, managers, and support personnel about electrical hazards and exposure controls and for understanding the electrical safety program so that everyone can contribute to its development and implementation. Managers, supervisors, and workers need to understand the program's structure, plans, and procedures. Having this knowledge ensures that everyone can fully participate in developing, implementing, and improving the program.

Additional training may be needed depending on the roles assigned in the program. For example, employers, managers, and supervisors may need specific training to ensure they can fulfill their roles in providing leadership, direction, and resources for the electrical safety program. Workers assigned specific roles in the program (e.g., incident investigation team members) may need the training to ensure their full participation in those functions. Employers, managers, and supervisors are responsible for workers' safety, yet sometimes have little training on safety-related concepts and techniques such as risk assessment, hierarchy of controls, and PtD. They might benefit from specific training on these, and other topics will allow them to fulfill their leadership roles in the program better.

Worker training should be tailored to the specific hazards and risks associated with their job responsibilities. Office workers with exposure to cord-powered office equipment should have training appropriate for this risk. Workers using mobile equipment or handling long conductive materials need training on the identification and avoidance of overhead power lines. Electricians whose job expectations include working near energized equipment and components need hazard identification and risk control training specific to their exposure to hazardous energy.

Additional training may be needed when a change in facilities, equipment, processes, materials, or work organization could increase hazards, and whenever a worker is assigned a new task. A formal process may be needed for the determination of the training needs of workers responsible for developing, implementing, and maintaining the program.

Design engineers may be familiar with minimum installation requirements in regulations, codes, and standards but not with methods to compare and evaluate the risk of various options. Design engineers may assume it is acceptable to rely solely on administrative controls during the construction and operation of a facility.

Promoting awareness of home electrical safety is a way to engage everyone in the organization in thinking about electrical safety. Management, supervision, and workers can apply home electrical safety tips in their personal lives. Topics can include GFCIs and arc fault circuit interrupters (AFCIs), extension cords, electrically powered tools and appliances, and overhead and underground utility power lines. Home electrical safety education can provide a sound base for understanding and value for the electrical safety program at work. The Electrical Safety Foundation International has an extensive library of home electrical safety awareness materials that can be downloaded at no cost.

Program Evaluation and Improvement

Once an electrical safety program is established, it should be evaluated initially to verify if it is being implemented as intended. After that, management should periodically, and at least annually, step back and assess what is working and what is not, and whether the program is on track to achieve its goals. Whenever these assessments identify opportunities to improve the program, employers, managers, and supervisors – in coordination with workers – should make adjustments and monitor how well the program performs as a result. Sharing the results of monitoring and evaluation within the workplace, and celebrating successes, will help drive further improvement.

The first step in monitoring is to define indicators that will help track performance and progress. A balanced scorecard of both lagging and leading indicators should be used. Lagging indicators track worker exposures and injuries that have already occurred. Leading indicators track how well various aspects of the program have been implemented and reflect steps taken to prevent incidents and injuries before they occur.

Electrical injuries comprise less than 0.2%, or less than 1 in 500, of all nonfatal occupational injuries. The relative infrequency of electrical injuries can create an illusion of having an effective electrical safety program. A rare serious electrical injury or fatality may be incorrectly viewed as a random event. Incidents and injuries that are low in frequency also can mean that a company or an organization does not have enough internal data to be statistically meaningful. Due to the inherent low-frequency occurrence of electrical injuries, an individual or a company may not recognize the potential for a fatal (high-consequence) injury. The lack of, or a low number of, electrical injuries is not a valid indicator of the quality of the electrical safety program. Factors that may be better indicators of electrical safety program quality are leading indicators such as:

- Frequency and quality of electrical safety training for electrical workers, non-electrical workers, line supervision, management, and support personnel such as safety professionals, contractor administrators, and training staff
- Frequency of field audits that examine the implementation of the organization's electrical safety program
- Quality and frequency of management system audits focused on preventing exposure to electrical hazards
- Attention to inherently safer design in hardware selection and electrical system design in capital projects
- Discipline in maintaining maintenance programs for equipment and systems critical to electrical safety.

At least annually, management should evaluate the program to ensure it is operating as intended, is effective in controlling identified hazards, and is making progress toward the established electrical safety goals and objectives. The scope and frequency of program evaluations will vary depending on changes in your organization, OSHA regulations, industry standards, and the scope, complexity, and maturity of your electrical safety program. The evaluation should take into consideration whether changes in equipment, facilities, materials, key personnel, or work practices trigger the need for changes in the program.

Safety management system standards provide the blueprint, or framework, to help enable effective, robust, and sustainable programs to manage occupational safety and health risks.

Before safety management systerms standards were developed, some companies developed proprietary safety management standards that aligned with or went beyond industry standards. The US Department of Defense published the first safety management standard, *Mil Standard 882, Standard for System Safety*, in 1969. The standard documented safety management practices that were developed to manage the extraordinary risk of advancements in aviation, nuclear weapons, and other advanced weapons systems. The first safety management standard focused on nonmilitary risks appeared in 1995, with the publication of ISO 14001, *Environmental Management Systems*. In 1999, a collaboration of international safety organizations published OHSAS 18001, *Occupational Safety and Health Management Standard*. A similar standard, *ILO Guidelines for Occupational Safety and Health Management Systems*, was published by the International Labour Organization in 2001. More recently, ANSI Z10, *Occupational Health and Safety Management Systems*, and CSA Z1000, *Occupational Health and Safety Management*, were first published in 2005 and 2006, respectively.

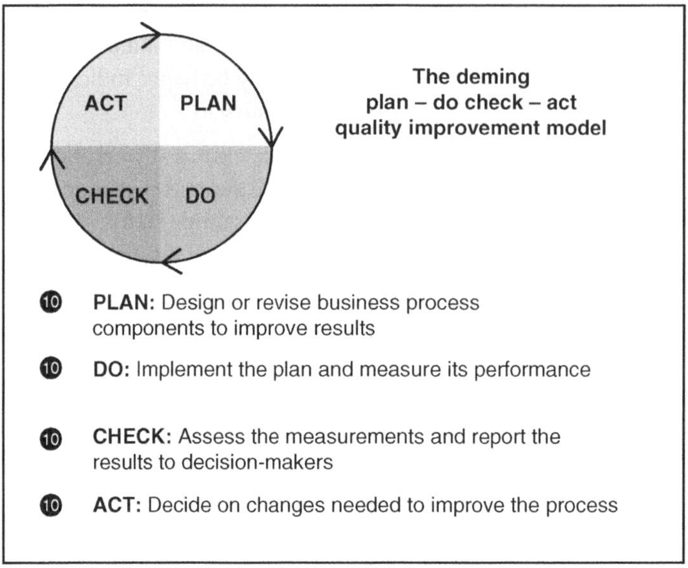

Figure 8.1 Elements of the Deming quality improvement model.

These safety management system standards are based on quality management principles attributed to W. Edwards Deming. The Deming quality improvement model *Plan-Do-Check-Act*, as shown in Figure 8.1, is central to the safety management standards. These standards are also well harmonized with the comprehensive hazard control measures commonly referred to as the thearchy of control measures or risk treatments. In addition, they are harmonized on how these equally important measures are ranked in descending order of relative effectiveness in helping assure worker safety. These aspects of safety management are not i discussed in detail in NFPA 70E. Recognizing this and understanding how to integrate the requirements of NFPA 70E within the framework of safety management systems can help an organization maximize the return on its investment in improving its electrical safety program.

The safety management system standards are based on proven principles that are fundamental and essential for robust safety programs and sustainable safety performance. While the strict implementation of requirements in OSHA regulations and NFPA 70E can enable an organization to realize performance improvement, optimum improvement is likely not achievable without integration with the proven strategies provided in the safety management system standards.

Overcoming Barriers for Small Organizations

The safety management standards noted earlier may be intimidating to small organizations that do not have safety management professionals on staff to facilitate implementation of the core elements of safety management. OSHA has two publications and a web-based toolbox of resources that can help any organization to design, implement, and continuously improve a safety management process uniquely tailored to the organization.

Effective application of safety management system standards typically requires having safety professionals competent in the standards. Many small businesses do not have the resources to have this competency in-house, and the task of understanding and implementing safety management system standards can be overwhelming. To aid small companies and other organizations lacking the resources needed to implement safety management system standards, OSHA has published guidelines to assist employers in the core elements of safety management proven essential for effective and robust safety programs. These core elements are interrelated and are best viewed as part of an integrated system. For electrical safety, this means any organization can now have easy access to resources to address the gaps in an electrical safety program not included in NFPA 70E. One of the resources is the OSHA Crosswalk of Voluntary Standards and Guidelines, which shows how the core elements in OSHA Recommended Practices for Safety and Health Programs are aligned with safety management system standards and leading organizations in occupational safety.

OSHA Recommended Practices for Safety and Health Programs should be tailored to the electrical safety program for your workplace. The core elements for successful safety management are common across all hazards, and the recommended practices and supporting resources can be applied to any safety and health hazard in the workplace. By inserting "electrical safety program" wherever "safety and health program" appears in the tools and resources, you will see opportunities to improve visibility, support, engagement, and robustness of your electrical safety program. Experimentation, evaluation, and program modification are part of the process. You also may experience setbacks from time to time. What is important is what you learn from setbacks, remain committed to finding out what works best for you, and continue to seek leading-edge methods.

Electrical injuries occur in all types of workplace settings, from manufacturing sites to hospitals and health care facilities, offices, construction, and to service industries. The preventive approaches described in the OSHA recommended practices work equally well across all sectors of the economy, for all different kinds of hazards, in both mobile and fixed work environments, and for small-, medium-sized, and large organizations. Small employers may find that they can best accomplish the actions outlined in these recommended practices using informal communications and procedures. Larger employers, who have more complex work processes, may require a more formal and detailed program. They also may wish to integrate their electrical safety program with other programs they are using to manage assets, production, quality control, and environmental protection or sustainability.

You can use the self-evaluation tools found on the OSHA Recommended Practices for Safety and Health Programs website to track your progress and assess how fully you have implemented each action item.

The Hierarchy of Controls

The application of a hierarchy of risk control measures is fundamental to safety management systems but is not incorporated directly in OSHA regulations for occupational electrical safety or in NFPA 70E. The application of a hierarchy of risk controls is a methodology used in safety management to minimize or eliminate exposure to hazards and to eliminate or reduce the risk of injury to ALARP. It is a widely accepted methodology with applications ranging from highly hazardous processes with potential for catastrophic consequences to life, property, and the environment to more common hazards, such as the use of hand tools.

The concept of applying a hierarchy of risk control measures emerged during the development of systems safety engineering that began in the 1940s as aviation technology evolved rapidly. Airplanes were becoming more complex and powerful machines. Mishaps were becoming more costly, and hazards to pilots were increasing. System safety engineering was the culmination of advancing knowledge in safety designs, equipment reliability, human error, and management systems that held the promise of preventing unacceptable or catastrophic loss. Demonstrated results from the application of system safety engineering in aviation safety led to applications in nuclear weapons, nuclear power generation, and highly hazardous petrochemical processes. The first safety management system standard to incorporate the concept of hierarchy of risk control measures was MIL STD 882, Standard Practice for System Safety, published in 1969. Since then, industry consensus standards, including OHSAS 18001, ANSI/ASSP Z10, CSA Z1000, and ISO 45001, have incorporated the concept of hierarchy of risk control measures. A hierarchy of controls provides a systematic way of analyzing hazards and risks and choosing the most effective means to eliminate or reduce hazards and their associated risks. Some of the literature in safety management systems describes the hierarchy as five

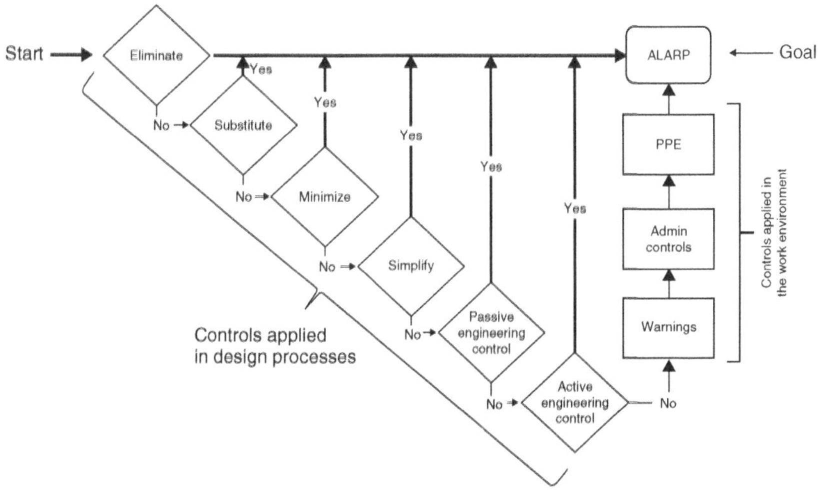

Figure 8.2 The hierarchy of controls. Adapted from Lyons and Popov and ANSI/ASSP Z10-2019.

elements, and sometimes as nine elements. This difference in describing the hierarchy of risk control measures is not significant and is usually due to the granularity of describing the controls. The concept of a hierarchy of risk controls has been evolving for more than 50 years. The model in Figure 8.2, based on Lyons and Popov's work, expands the granularity for risk control solutions. Risk treatments on the diagonal axis from the upper left to lower right are applied in the design and redesign processes. The risk treatments on the vertical axis are applied in the Workplace to risks that were not reduced to ALARP in design. The first four treatments, elimination, substitution, minimization, and simplification, are long-lasting but may require maintenance to ensure effectiveness. Any remaining risk will not change over time unless the design is changed. Passive and active engineering controls must be combined with inspection, testing, and maintenance (administrative controls) to ensure functional integrity. Improper maintenance can result in hidden failures of engineering controls.

In applying a hierarchy of controls, the outcome should be that the risks for which the probability of an incident or exposure occurring and the severity of harm that could result are ALARP. If the risk of injury cannot be reduced to an acceptable level in design, then the controls shown vertically on the right must be considered. Risk controls applied in the work environment are considered less effective than those used in design processes due to vulnerability to human error or to deliberate decisions by management or workers at risk. The top three control measures, elimination, substitution, and engineering controls, are more effective because they result in a more inherently safe installation, are less dependent on error-free performance of supervision, support personnel, and the workers at risk, and are less vulnerable to being defeated by supervisors or workers. The bottom three control measures, warnings, administrative controls, and PPE, are vulnerable to human error and are highly dependent on the performance of supervision, support personnel, and workers at risk. Each of the categories of risk control measures is discussed in the subsequent sections.

Elimination

If hazards are eliminated in design, then the risks derived from those hazards are also eliminated. Although rarely feasible, permanent elimination of hazardous electrical energy may be possible. Consider the example of the design of a picnic shelter for a city park. The original design included electric lights and convenience outlets. An early design review noted that the park closed at dusk and opened at dawn, so the electric lighting and outlets were deleted. This design change permanently eliminated the hazard and risk of electrical injury for the life of the picnic shelter.

We live in an electrical world with nearly every aspect of our modern society dependent on electrical technology, with its inherent risk of injury. In practice, the elimination of a hazard completely by modifying the design is usually not practicable. More often, the goal is to modify the design so that the likelihood of human errors and the need for PPE is at a practical minimum. An example is to locate overhead electric lines away from vehicular traffic roadways, material storage and handling areas, and areas where cranes or other mobile lifting equipment may be required for construction or maintenance activities. This example is not pure elimination, as the hazard is still there, but additional space was created to better isolate work activity from the overhead lines.

Substitution

Substituting less hazardous equipment, materials, or energy can result in reducing the frequency or potential severity of exposure. Substituting a 24-V control in the design of a process is an example of selecting a less hazardous energy. Selecting arc-resistant switchgear in lieu of non-arc-resistant switchgear is an example of selecting less hazardous equipment.

Minimization Minimization is a subset of substitution. Can the initial design be modified by reducing the number of motors or lighting fixtures? Minimizing the number of electrical components in the design will likely reduce exposure in operations and maintenance.

Simplification Simplification is also a subset of substitution. Modifying a design to reduce the number of lockout points to isolate a section of switchgear for maintenance is an example of substitution.

Engineering Controls – Passive Passive engineering controls are design choices that function automatically, without any action by personnel, to reduce the frequency or consequences of exposure to a hazard. A permanently installed GFCI is an example of an engineering control that automatically reduces the consequence or severity of an electric shock exposure. A worker plugging in a cord-powered tool does not have to take any action to enable the safety benefit of the GFCI.

Engineering Controls – Active On the other hand, an active engineering control requires action on the person at risk of injury in order to achieve the safety benefit of the control. A portable GFCI is different from a permanently installed GFCI as the person at risk must take action to use the portable GFCI.

Warnings

Warnings may be temporary or permanent, and audible or visible. Signs and labels, lights, barriers and barricades, and alerting personnel are examples of warnings. Warnings are highly dependent on administrative controls such as training, maintenance of legibility, and visibility. The effectiveness of warnings is vulnerable to human error in understanding and response.

Administrative Controls

Administrative controls include safe work practices, standard operating procedures, maintenance programs, personnel selection, training, work scheduling, permitting systems, audits, and management of change. The creation of an electrically safe work condition includes a combination of administrative controls/safe work practices. This combination includes hazard identification, energy isolation, draining of stored energy, and isolation verification by visual inspection and with voltage testing instruments. For many organizations, it is common that administrative controls comprise the primary approach to risk management. The effectiveness of administrative controls is highly dependent on human performance and operational discipline of supervision and support personnel throughout the organization. The worker at risk of injury is dependent on administrative controls being properly designed, maintained, and implemented by other personnel as well as the worker's own knowledge, experience, and operational discipline.

Personal Protective Equipment

Electrical PPE includes clothing and accessories to minimize the severity of injury when there is potential exposure to hazardous electrical energy.

The proper use of PPE relies heavily on a series of administrative controls that include hazard assessment, selection, fitting, training, inspection, and maintenance to help ensure that the PPE is available when needed, the worker recognizes the need, and it is used properly. The use of PPE is considered the least effective control measure because of the limitations of administrative controls noted earlier, as well as potential errors in hazard recognition and errors in proper selection and use by the worker at risk.

All PPE should be safely designed and constructed and be maintained in a clean and reliable manner. It should fit comfortably, encouraging worker use. If the PPE does not fit properly, it can make the difference between being safely covered and dangerously exposed. When engineering, work practice, and administrative controls are not feasible or do not provide sufficient protection, employers must provide PPE to their workers and ensure its proper use. Employers are also required to train each worker required to use PPE to know:

- When it is necessary
- What kind is necessary
- How to properly put it on, adjust, wear, and take it off
- The limitations of the equipment
- Proper care, maintenance, useful life, and disposal of the equipment

A Challenge to Electrical Safety Experts

Electrical engineers, electricians, and safety professionals involved in designing and implementing improvements to a workplace electrical safety program tend to focus on the technical aspects, including safe work practices, electrical system design, hazard analysis, and the performance of PPE. The fundamental foundations of safety management may be overlooked if the electrical safety program is delegated solely to the electrical experts. For someone with concern or responsibility for an organization's electrical safety program, exploring the answers to the following questions may hold the key to enabling success:

a) How familiar are you with ANSI Z10, *Occupational Health and Safety Management Systems*, and ISO 45001, *Occupational Health and Safety Management*, or other globally recognized safety management system standards?
b) How do these standards align with the safety management systems in your organization?

c) How effective have you been in selling electrical safety improvement initiatives to your management?
d) How would you rate the collaboration and synergy among the safety professional, key members of management, and the electrical experts with respect to driving improvement in the electrical safety program in your organization?

For those not fully comfortable with answers to these questions, a better understanding of the role of safety management systems in planning and implementing changes in an electrical safety program may be one of the most critical factors in the success of the electrical safety program, no matter what stage of implementation or its level of maturity is. An organization just beginning to apply the requirements of NFPA 70E, an organization that has a mature safety management system and electrical safety program, but has not assessed integration effectiveness will likely benefit from a critical review of its electrical safety program and safety management systems. As illustrated in the case history in Chapter 4, one company with a mature safety management system and a mature electrical safety program demonstrated breakthrough and sustainable performance in electrical fatality prevention by undertaking a critical review of the electrical safety program integration into its overall OHS management processes.

This integration can help improve the effectiveness, robustness, and sustainability of the electrical safety program and better help realize the benefits of the electrical safety program as an asset critical to operational excellence, as opposed to a cost to business operations.

The recognized safety management system standards are from different standards development organizations, but are well aligned and based on proven quality management principles. Electrical engineers and specialists may not be experts in safety management systems. Safety professionals may not be experts in electrical safety. Top and middle management control priority, budgetary, and staffing decisions that are critical to the effectiveness of electrical safety program. An organization intent to achieve real improvement in its electrical safety program should consider organizing a collaborative effort involving electrical subject matter experts, safety professionals knowledgeable in safety management systems, and key members of management. Collectively, these skills, knowledge, and responsibilities can create an extraordinary collaboration and synergy to assess and improve your electrical safety program.

Electrical safety programs could be greatly enhanced if stakeholders in electrical safety nurtured the following key points:

- Promoting discussion of safety management systems, including the concepts of continuous improvement and comprehensive hazard control measures provided by widely recognized safety management system standards.

- Fostering a framework for creating synergy and collaboration among electrical subject matter experts, safety professionals knowledgeable in safety management systems, and management having responsibility for business integration, goal setting, priority resolution, and resource allocation.
- Encouraging enhanced or inherently safer engineering solutions in equipment and system designs that serve to eliminate or reduce the severity and frequency of exposure and risks from electric shock and arc flash hazards.

Collectively, these skills, knowledge, and responsibilities can create an extraordinary collaboration and synergy to assess and improve your electrical safety program. Examples of initiatives that could emerge from this collaboration include:

- Development and implementation of a corporate Health, Safety & Environment (HSE) electrical safety standard to replace a corporate engineering electrical safety standard. This standard applies globally to all operations.
- Development and implementation of second-party electrical safety audit, integrated into the corporate HSE auditing process.
- Inclusion of electrical safety needs in the corporate and business unit HSE budgeting processes.
- Inclusion of electrical safety resource staffing in the corporate and business unit HSE planning processes.
- Inclusion of the electrical safety program quality review in corporate and business unit HSE management processes.

Crosstalk: Questions for Discussion

1. Does your organization have an OSHMS to ensure continuous improvement in identifying and controlling hazards in its operations?

2. How does management communicate and support its commitment to electrical safety?

3. What functions in your organization are involved in managing your electrical safety program? Does the responsibility for managing the electrical safety program reside with just one function, i.e. electrical maintenance, engineering, or safety?

4. If you were asked to plan the annual management review of the electrical safety program, what would the review include?

5. Who has the responsibility for educating management on issues critical to electrical safety?

Further Reading

ANSI Z10, *Occupational Health and Safety Management Systems, American Society of Safety Professionals.*

Driscoll, T.S., *Owners Implementation of Safety Programs*, 2008 *IEEE IAS Electrical Safety, Technical and Mega Projects Workshop, November 19–21, 2008*, Edmonton, Alberta.

Floyd, H.L., *Operational Excellence and the Impact of an Effective Electrical Safety Program*, 2008 *IEEE IAS ESTMP Workshop, November 19–21, 2008*, Edmonton, Alberta.

Floyd, H.L., The Potential Impact of Safety Management Standards, *IEEE Industry Applications Magazine*, May/June 2011.

ILO-OSH, *Guidelines on Occupational Safety and Health Management Systems*, International Labour Organization, Geneva, Switzerland.

ISO 14001, *Environmental Management System*, International Standards Organization, Geneva, Switzerland.

Mil Std 882, *Standard for System Safety*, US Dept if Defense.

NFPA 70E, *Standard for Electrical Safety in the Workplace*, National Fire Protection Association, Quincy, *MA*.

OHSAS-18001, *Occupational Safety and Health Management Systems – Requirements*, The Occupational Health & Safety Group, Cheshire, UK.

OSHA 3885, *Recommendations for Safety and Health Programs.*

OSHA 3886, *Recommendations for Safety and Health Programs in Construction.*

Tompkins, P. (2025). Safety management system planning. *Incident Prevention* 22 (3): 14–15.

9

Prevention Through Design

Modern awareness of workplace hazards has created a range of safety standards and best practices now accepted and implemented in the design and construction of many types of industrial and commercial installations. PtD concepts are well understood and commonly applied in the design of many facilities where dangerous chemicals or dangerous concentrations of energy are a required part of the process. Heightened awareness of electrical hazards has resulted in a greater understanding of the arc flash hazard and shock hazard and in many methods and tools to control the associated risks. However, the PtD concepts, though well understood and implemented in other aspects of industrial facility design, may not be well understood or are often not implemented at the electrical infrastructure's design and construction stage within industrial or commercial facilities. This chapter discusses the correlation between PtD and industrial accidents and how PtD concepts applied to electrical infrastructure can improve modern facilities' safety and productivity.

Though the workplace in modern economies is significantly safer than the workplace of a few decades ago, workplace hazards remain. Electrical injuries and fatalities in developing countries are increasing, as the expansion on the application of electrical technologies exceeds the pace of efforts to manage risk. Designing systems, processes, and facilities to improve the safety of those who must operate and maintain facilities is generally a well-known art based on well-understood principles of identifying hazards and controlling risks. For example, the "Hierarchy of Risk Controls" initially documented in US MIL-STD 882 Standard Practice for System Safety in 1969 is now widely accepted as a core concept in safety management.

The Value of Prevention

Industrial accidents in modern society can have very high costs: loss of life, costs associated with injury recovery, costs related to loss and replacement of

Hidden Risk in Occupational Electrical Safety: Through the Lens of System Safety,
First Edition. H. Landis "Lanny" Floyd.

Table 9.1 Examples of industrial disasters having significant implications to people and environment.

Texas City, USA (1946)	Port explosion. 699 fatalities, 3,500 people injured, 1,519 houses destroyed
Flixborough, England (1974)	Chemical plant explosion. 28 fatalities, 89 people injured, 1,500 surrounding buildings damaged.
Seveso, Italy (1976)	Dioxin release. 3,300 farm animals suffered fatal exposure; 80,000 animals slaughtered to prevent dioxin from entering food supply; 220,000 people monitored for health effects.
Three Mile Island, USA (1979)	No direct injuries to people, halted expansion of nuclear power generation in the United States.
Chernobyl, USSR (1986)	Nuclear power facility fire and explosion. 56 fatalities onsite, 350,000 people relocated.
Bhopal, India (1988)	Toxic chemical leak. Fatality estimates exceed 15,000; injury estimates exceed 500,000.
Piper Alpha, North Sea (1988)	Offshore oil platform explosion, impacting 10% of North Sea gas production. 167 workers killed.
Texas City, USA (2005)	Refinery explosion. 15 fatalities, 170+ injuries
Deepwater Horizon, Gulf of Mexico (2010)	Offshore oil platform explosion. 11 killed, 17 injured, largest offshore oil spill in US history.

assets, costs related to lost production, costs associated with workers' morale and productivity, and costs associated with negative publicity. A limited list of high loss examples is provided in Table 9.1, with significant categories of financial implications described in Table 9.2. Any modern facility where machinery, chemicals, or stored energy poses risks to workers or the environment will implement machinery, chemical, facility design, and environmental safeguards due to the need for statutory compliance, because it is economically logical to do so and because it is the right thing to do. Incident and accident rates must

Table 9.2 Cost sources for an incident.

Loss of production: Inability to manufacture, inventory spoilage, raw materials & energy wasted, contracts unfilled, and customers lost
Loss of capital assets: Replacement costs, cleanup cost, removal and reinstallation cost, equipment replacement cycle
Injury-associated costs: Medical, fines, worker replacement
Morale/psychological impact: Costs associated with worker dissatisfaction, fear, organized labor actions
Reputation: Wall Street impact, investor impact, client impact, supplier impact

be kept as low as possible to keep productivity as high as possible and be a good corporate citizen. Economic and safety statistics indicate that industrial safety and economic productivity are closely correlated. Which is a cause and which is effect may be more arguable, but studies by various organizations and parties demonstrate that productivity and safety are correlated throughout the world.

The sources and potential magnitude of cost for an enterprise will vary, but it is not difficult for the negative consequences to add up to millions of dollars for both events involving human injury and those that do not. They often cause lost production or damaged capital assets required for production. On occasion, they are reflected in front-page news, devaluations of stock in capital markets, and even insolvency. In addition, the causes of both safety hazards and extreme financial loses are often the same, thus ignoring what could happen. Hence, many investments thought of as safety related may also be considered to positively impact productivity and reliability, and many that may be made in the interest of productivity and reliability may have safety benefits.

Safety at the System Level

Electricity is a very valuable energy delivery and usage mechanism. None of us lead an electricity-free life in our personal environments or our professional environment. Electricity is used more safely today than it was decades ago. Our homes are full of innovations introduced over time that improve consumer electrical-related safety, such as residual current devices (RCDs), GFCI protection, AFCI protection, double-insulated tools, and more. Installations in the United States, residential, commercial, and industrial, regulated by the National Electrical Code (NEC), which is updated every three years, and local codes and industry-specific codes and standards, are much safer today than just a few decades ago. NFPA 70E provides guidance to properly implement OSHA regulations associated with electrical work in places of employment. The design features of modern tools, appliances, and other cord-powered household devices are inherently safer due to PtD applications that we now take for granted. The same rigor may not be applied to preventing the opportunity for electrically related "mishaps" and reducing "residual risk" associated with the use of electrical energy throughout the life of an industrial or commercial power system.

MIL-STD-882 states: *system safety is the application of engineering and management principles, criteria, and techniques to achieve acceptable mishap risk, within the constraints of operation, effectiveness, and suitability, time and cost, throughout all phases of the system life cycles.* This standard is written from the perspective of military applications. However, the principle stated here may be considered universally applicable, where "acceptable mishap risk," also known as residual risk, is very low. Residual risk may be expressed as:

Residual risk = the risk remaining after risk reduction measures have been applied

Few activities result in zero risk; however, at some point, residual risk is determined to be so low as to be acceptable to continue the activity. We all drive cars or take public transportation at one time or another. But there should be no delusion that doing so is entirely risk free. The Hierarchy of Risk Controls is implemented every time we get in a vehicle; we make the conscious or unconscious decision that the "substitutions" such as a fully enclosed automobile instead of an open motorcycle; the "engineered controls" such as good brakes and traffic control systems; the "awareness" provided by proper roadway signage; the "administrative controls" provided by driver's training; and the "PPE" offered by seat belts and airbags reduce the inherent risk. These controls reduce residual risk to one we are willing to accept to go shopping, go to work, pick up the kids at school, or drive to the airport where we aboard another conveyance where we hope the Hierarchy of Risk Controls was particularly well implemented. It is good to note that we did not voluntarily select many of these hazard controls we now take for granted. They were driven by governmental codification of safety requirements similar to how the National Electrical Code (NEC) has codified GFCI use, U.L. has standardized wall outlets and other safety-related, now commonly accepted, electrical practices. For example, consider the insertion and removal of the plug on a cord-powered tool or appliance into a wall outlet. Patented by Harvey Hubbell in 1913, the touch safe design of the outlet and guarding design of modern plugs reduce the risk of unintentional contact with energized parts to acceptable levels, without the need to use voltage-rated gloves when making a temporary connection to an energized power source.

Risk Assessment and Control

Risk management and the Hierarchy of Risk Controls are gaining acceptance in the electrical community. HSE professionals are familiar with the concepts and apply them to task planning, training, work plans, and PPE selection on a routine basis. However, the same concepts can also apply to the system design process, focusing on the higher order controls within the hierarchy of controls. Within the design process, risk can be examined from the hazard or task perspective. Can a hazard be addressed via elimination, substitution, or engineering controls, or can a hazardous task be managed via the same methods? If a hazard cannot be removed, removing the need for the activity may be almost as impactful.

Risk assessment includes the following processes:

- **Identification of the hazards or risks** inherent in a design. According to Ericson, hazard identification includes considering energy sources, equipment function, and operational tasks. Concentrations or sources of energy are potential hazards. From substations (electrical energy) to anything that can fall, including the worker (potential kinetic energy), are hazards and tasks that require the worker to be near or interact with the hazard in such a manner that can allow an unplanned release of the energy to harm the worker is a potentially hazardous activity. Can a hazard be eliminated or diminished?
- **Risk analysis and estimation:** Are hazards large or small, unacceptable, or tolerable; incidents probable and predictable or rare and easily preventable? Are dangerous tasks to be performed rarely under very controlled circumstances or often under time and production pressure? Are potential consequences severe or minor? For example, higher voltage versus lower voltage levels or high potential arc energy per unit time versus low energy per unit time are easily identifiable and, to some degree, measurable. The probability of an incident may be harder to assess numerically; however, comparative assessments may be enough to guide decisions. For example, are incidents causing an accidental electrical arc more prevalent in solidly grounded low-voltage (LV) systems or in impedance-grounded LV systems? Will lights need to be turned on/off often may help decide if it is best to have a dedicated light switch or use a CB within a lighting panel to operate the light.
- **Risk evaluation** could be the point at which whether the residual risk or hazard that remains after effecting risk control is acceptable. Acceptable in a design context does not mean so low as to no longer need risk controls, but low enough that it can be left up to the implementation of this same process in the field to handle the residual risk via the lower order controls of the Hierarchy of Risk Controls.

Risk Treatment or Control

Brief descriptions and examples of each risk treatment strategy in the hierarchy of controls discussed in Chapter 8 are as follows:

1) Eliminate. Hazards/risks are eliminated or removed from systems/workplaces during design reviews. Elimination is achieved when there is no foreseeable way a person could be exposed to the hazard under assessment. It may be rare in our electrical world to identify examples in which a specific electrical hazard can be permanently deleted in design. However, there may be some

situations in which electricity or a specific item of electrical equipment could be permanently deleted during design.

2) Substitute. Identified hazards energy is intentionally replaced with less hazardous energy, for example, using 24-V controls instead of 120-V control.
3) Minimize. The number of devices having a particular hazard is minimized to present a lower frequency of exposure.
4) Simplify. The likelihood of error or occurrence is reduced by reducing system or workplace complexity.
5) Passive engineering controls. Risks are controlled or contained without any action required by the person at risk.
6) Active engineering controls. Risk reduction is achieved by devices requiring action by the person at risk.
7) Warnings. Signs and labels, lights, barriers and barricades, and alerting personnel are examples of warnings. Warnings are highly dependent on administrative controls, such as training, maintenance of legibility, and visibility. The effectiveness of warnings is vulnerable to human error in understanding and response.
8) Administrative controls. Administrative controls include safe work practices, standard operating procedures, maintenance programs, personnel selection, training, work scheduling, permitting systems, audits, and management of change. The creation of an electrically safe work condition includes a combination of administrative controls/safe work practices. This combination includes hazard identification, energy isolation, draining of stored energy, and isolation verification by visual inspection and with voltage testing instruments. For many organizations, it is common that administrative controls comprise the primary approach to risk management. The effectiveness of administrative controls is highly dependent on human performance and operational discipline of supervision and support personnel throughout the organization. The worker at risk of injury depends on administrative controls being properly designed, maintained, and implemented by other personnel and the worker's knowledge, experience, and operational discipline.
9) PPE. The proper use of PPE relies heavily on a series of administrative controls that include hazard assessment, selection, fitting, training, inspection, and maintenance to ensure that the PPE is available when needed, the worker recognizes the need, and the PPE is used properly. The use of PPE is considered the least effective control measure because of the limitations of administrative controls noted earlier, as well as potential errors in hazard recognition and errors in proper selection and use by the worker at risk.

The total available energy in a power circuit can be reduced with a smaller transformer or advanced circuit protection that limits incident energy. However, other actions may not act on the hazard as much as they act upon the activity that, combined with the hazard, creates risk. Automation, such as remote operation or remote human–machine interface (HMI) with the ability to provide measurement, "useful" diagnostics, or control, minimizes near-human interaction and decreases both exposure and probability of human error. When selecting characteristics for equipment selection or equipment performance, it is necessary to understand how the capability may be used to lower risk to maintenance and operating personnel, as well as to understand how it brings value to the overall operation of the facility. Can voltage be reduced below hazardous levels? Consider how many lumens can be obtained from 24 V lights today. Is 120 V or 277 V needed for lighting today as much as 20 years ago? Can a hazard be eliminated by availing the design of the latest technology instead of traditional methods?

Consider how much interaction with equipment is required to perform primary current injection testing of CBs. Is it as necessary today as it was 20 years ago? Can modern CBs designed with electronic trips with powerful embedded processors, communications, and advanced sensing provide information over time, allowing a user to determine the device's condition?

If exposures inherent in maintenance practices can be decreased while improving equipment reliability and safety, then the hazard may not be affected, but the frequency or duration of the hazardous activity may be reduced. The net effect is that the risk is reduced to those who will operate and maintain the system.

Residual Risk

When addressing workplace risk, it is often not possible to achieve a zero-risk situation. When implementing PtD, it may not be possible to eliminate a hazard with higher order controls. However, it may also be possible to replace one hard-to-control hazard with one that is more easily managed via lower order risk controls implemented by or for the worker. Furthermore, an acceptable residual risk in one situation may not be a level of risk that may be deemed acceptable in another. Consider the risk associated with locating a 600-A 480-V disconnect switch in an industrial facility's lunchroom. Although the switch is a common piece of electrical equipment in an industrial facility, locating the switch in a space that would be routinely occupied by workers who may not know the risk of arc flash injury during normal and servicing situations creates risk to those workers. Locating the switch in a space reserved for electric power equipment and accessible only by qualified workers would isolate the arc flash hazard from lunchroom occupants. Although the arc flash hazard is the same in both locations, the residual risk to workers is different.

Risk is usually defined as a combination of the existence of a hazard and the probability of human exposure to the hazard. If exposure is possible, then risk controls must be applied to achieve an acceptable level of risk. But the likelihood of the event can also be divided into the opportunity for the incident, i.e., man-hours of exposure, and the probability during the exposure. If the need for a task can be eliminated, then it could be considered akin to the elimination of exposure to a hazard. If the need for the task can be diminished or replaced with a less hazardous task, then the action can be considered akin to a substitution or engineered control aimed at a specific hazardous task. If the task or environment for the task is changed such that the probability of an accident is less, the overall risk is reduced. High resistance grounding (HRG) versus solidly grounded systems, remote operation versus close operation, supervision of automatic switching versus manual switching are all design elements that address event probability. All else being equal, an HRG system is expected to be safer because the possibility of an arcing ground fault is near zero, remote operation keeps operators from components with a nonzero probability of failure, and automatic switching should be less subject to human error. However, in many cases, there may be some residual risk that cannot be addressed by a factor under the designer's control or inherent to the activity. What is acceptable at this point will vary with the context of the situation.

Prevention Through Design

The US NIOSH leads a US national initiative called Prevention through Design (PtD). The purpose of the initiative is to promote the concept that one of the best ways to prevent and control occupational injuries, illnesses, and fatalities is to "design out" or minimize hazards and risks in any project or system. Countries of the EU, Great Britain, and other countries that have long embraced a risk-based safety culture, including PtD, have demonstrated significant improvement in fatality reduction compared to the United States. The PtD initiative, as defined by NIOSH, includes the following:

- Permanently eliminating hazards and controlling risks to workers to an acceptable level "at the source," means questioning if the equipment or circuit under review can be permanently deleted from the design
- Permanently eliminating hazards and controlling risks to workers to an acceptable level as early as possible in the life cycle of items or workplaces
- Including design, redesign, and retrofit of new and existing work premises, structures, tools, facilities, equipment, machinery, products, substances, work processes, and the organization of work
- Enhancing the work environment through the inclusion of prevention methods in all designs that impact workers and others on the premises.

PtD is a collaborative initiative that is based on the principle that the best way to prevent occupational injuries, illnesses, and fatalities is to anticipate and "design out" or minimize hazards and risks when new equipment, processes, and business practices are developed" and emphasized that the PtD process requires cross-disciplinary activities.

PtD encompasses all the possible efforts to anticipate and design out hazards to workers in facilities, work methods and operations, processes, equipment, tools, products, new technologies, and the organization of work. PtD application in the design phase can impact the risk of electrical injury to workers throughout the operational life cycle, including construction, operation, maintenance, and eventual demolition. The initiative, as defined by NIOSH, includes stakeholder input as an initial step to research, education, practice, and policy to develop strategic planning that results in the implementation of occupational safety and health considerations within system design with a stated goal of preventing occupational injuries, illness, and fatalities. Guidance for practical application of PtD is provided in ANSI/ASSP Standard Z590.3-2019, *Prevention through Design Guidelines for Addressing Occupational Hazards and Risks in Design and Redesign Processes.*

PtD must be accomplished not only by examining, questioning, and addressing the hazards that may be present in any system but also by reviewing, questioning, and addressing any expected activity that may bring workers in proximity to those hazards and/or require interaction that puts workers in harm's way. Risk analysis may be performed from the perspective of hazards or tasks that expose the workers to hazards. From the perspective of PtD, both reducing the first and minimizing the second have value.

The System Life Cycle

The phases of the designed system life cycle may be considered as follows:

- **Conceptual design:** Where the high-level decisions, such as energy sources, are selected
- **Detail design:** Where detailed decisions are made, such as what type of equipment to specify, how to distribute power from source to application point
- **Contract assignation:** When contractors and suppliers are selected and contracts assigned. Often, a time and process step where previously decided design details can be compromised due to real or perceived budgetary or schedule limitations or goals
- **Construction and installation, execution:** Details may be added or subtracted to enhance or detract from future safety, and during this stage, hazards and risks may first appear, putting construction and installation workers at risk.
- **Operation:** The normal productive operation of the facility and process within the facility

- **Maintenance:** The upkeep, cleaning, maintenance, repair, and troubleshooting required to keep machinery, systems, and the overall facility in proper operational order
- **Demolition:** The eventual need to shut down and properly dispose of the facility, machinery, and residual elements of the process

Elimination and substitution, being the more preferred ways to control risks, usually depend on decisions taken early in the design process, such as whether to use difficult-to-maintain group-mounted equipment or easy-to-maintain compartmentalized draw-out switchgear, to use metal-clad versus metal-enclosed medium-voltage equipment, or whether to implement robust communications and sensing to maximize the ease with which future remote operations or condition-based maintenance may be implemented. As the design progresses, detailed decisions on circuit size, devices, protection packages, and protection capabilities are designed into the project to implement engineering controls to further control risk. Eventually, the design engineering team, in conjunction with the future maintenance and operation teams, can examine the expected operational, maintenance, and training requirements as well as the expected implementation of lower order risk controls.

Suppose residual risks are not sufficiently controllable or the lower order controls required to achieve acceptable residual risk are too complex, onerous, expensive, or unreliable. In that case, previous design decisions may need to be reevaluated to further control hazards with higher value, higher order controls to reach more desirable residual risks, allowing implementation of reasonable and acceptable lower order controls. If a process such as this is not followed, by the time an operator or maintenance team realizes that risks cannot be sufficiently controlled or require onerous maintenance practices, it may be too late to correct the situation cost-effectively.

All human activity is prone to some level of human error. If that activity occurs near or requires interaction with a dangerous energy level, human error can cause a hazardous release of that energy, regardless of the efforts to minimize error and control energy levels. Furthermore, human interaction with equipment can create future and hidden hazards for others or lower the reliability of that equipment due to human error. One approach to reducing the possibility of human error impacting system reliability and creating hazards is to minimize the activity. One way to accomplish that may be to implement sensing, automatic actions, or controls that supplant human activity. In many cases, automation can reduce the potential for human error and improve reliability and productivity.

Many electrical maintenance activities consist of measurements of one type or another. If you consider the maintenance activity around CBs or switchgear, much of it consists of measurements, tests, or what could be classified as monitoring for potential failures or indications of potential miss-operation.

The need to effect repairs is infrequent in devices that are not placed in harsh atmospheres, or are not subject to wear due to constant motion, such as motors.

For example, an IR thermal survey requires a person to observe, via appropriate instrumentation, all potential hot spots in equipment. Usually, field-installed electrical connections, moving contact assemblies, or connections that depend on spring pressure are considered likely failure points. The risk associated with the activity can be diminished by installing IR windows in the equipment. However, it may be diminished more, and the quality of the data gathered significantly improved by installing permanent thermal sensing that can provide data continuously. Correlating the thermal data with simultaneously gathered current values can increase the value of the information. Such information can be used to sense high resistance connections before high temperatures are reached and drive focused maintenance activity that exposes workers for much less time, therefore increasing system reliability. The key is in how the data is gathered, communicated, and processed.

CBs, sensors, and circuits could potentially yield data about the subsystems that can be correlated with operational data to determine wear, subsystem viability, wiring conditions, and even mechanical subsystem conditions, which otherwise would require manual maintenance. In some cases, an automated method may not yield all the desired information; however, if a high enough portion of desired data is obtained via automated techniques, the frequency with which the remaining information is manually procured can be considered for reduction, particularly if the remaining information is associated with subsystems less prone to failure.

"Other" Workers

For many organizations, the focus and improvement in electrical safety programs have primarily addressed electrical workers whose work tasks involve construction, operation, and electrical equipment maintenance. However, nearly 50% of workplace electrical injuries and fatalities are not electrical workers. Managers and administrators, painters, truck drivers, farmworkers, groundkeepers, and gardeners are among the top 10 occupations having the majority of fatal electrical injuries. For these nonelectrical workers or "other" workers, the exposure to electrical hazards ranges from the use of cord-powered portable tools and appliances to unintentional contact with overhead power lines in the course of routine work activities. When assessing designs for PtD applications, consideration should not be limited to electrical workers and should include all personnel who may be potentially exposed to hazardous electrical energy.

A Special Case: Construction

Electrical safety is a concern for all segments of the workforce. However, the construction industry may provide the greatest opportunity for improvement through the application of PtD concepts. As noted by NIOSH researchers Cawley and Homce, the construction industry accounts for 7% of the US workforce, but suffers 45% of electrocution fatalities. It is the author's observation that when electrical safety is addressed in design, the targeted work exposures are those of the operating and maintenance workforce. The safety control measures to protect the construction worker during the installation and demolition phases of an installation are not normally addressed in a similar manner. In many cases, the construction workforce is left with little choice but to perform work tasks where there is exposure to electrical hazards. This occurs so often that construction workers believe this is "part of the job" of being a construction worker. When construction personnel are invited to participate with design personnel and discuss the design, too many times the construction personnel take the attitude that it can be "fixed in the field" and will agree to accept a more hazardous design. This concept must be changed so that construction work forces will begin to recognize how things can be changed to reduce or eliminate the exposure to electrical hazards. Construction personnel need to provide recommendations back to design organizations on how to reduce these exposures.

Learning from Other Countries

In the 1980s, the European Commission performed a study indicating that 60% of fatal accidents in construction had a contributing factor related to an action or decision prior to the start of construction. This study led to the 1992 development of the EU Directive on Temporary or Mobile Worksites. The Directive placed legal responsibilities on project owners and on those associated with the design of the work. The Directive required member countries to put a requirement in place to meet the objectives of the Directive within five years. In 1994, the United Kingdom developed and implemented the Construction (Design and Management) Regulations (CDM). These regulations required construction companies, project owners, and architects to address safety and health during the design phase of projects, and companies to have responded with positive changes in management practices to comply with the regulations. The Australian government developed the Australian National Occupational Health and Safety Strategy 2002–2012, which set "eliminating hazards at the design stage" as one of five national priorities. The Australian Occupational Health and Safety Act has a requirement called "Duty of Care." This places requirements on those who have influence on the hazards in a workplace, including employers, employees, design organizations, and suppliers of equipment and materials. The intent is to require all to use "Reasonably Practical" means to identify the hazards and provide solutions within the constraints of the business environment. As a duty holder, you must show that you have exercised "due diligence" in providing a safe working environment.

Opportunities

Construction personnel need to be more involved in the early stages of the design of a project. As demonstrated in the 2007 PtD Workshop, there is a great desire by construction personnel to be additive in the reviewing design and commenting on how to make the installation safer. Rarely are construction personnel allowed to be involved early enough to have an effective impact. Design decisions made early are difficult and expensive to change later in the project. Construction participation in the design requires the owners of the facility to support the concept. Construction management organizations that represent the owner must also buy into the concept by providing knowledgeable personnel to participate with the design organization when the actual installation contractor has not been identified.

Some examples of how design can impact the exposure to electrical hazards are as follows:

- Indicating on construction drawings the location of underground lines or electrical lines embedded in concrete floors
- Not locating equipment, structures, or storage yards near or under existing overhead lines
- Designing temporary construction power systems to reduce electric shock and arc flash exposures and arc flash incident energy magnitudes
- Unitization of electrical systems so that energization can be scheduled without additional exposure, or minimal additional exposure, to continuing construction activities
- Having up-to-date drawings to facilitate demolition planning

What if facility owners, equipment manufacturers, and facility designers embraced an enhanced value for addressing the safety of the construction worker? What if there was an industry standard to guide the application of PtD tools and concepts that spanned the full life cycle of installations, from construction through commissioning, operation, maintenance, and demolition? The future presents the opportunity for different approaches.

Table 9.3 is a concept for a Construction Electrical Hazards Analysis planning tool that could be expanded or modified to aid in identifying and addressing electrical hazards early in the design planning of a construction project. As noted in the table, this checklist is not all inclusive and should be expanded through future

Table 9.3 Construction electrical hazards design analysis checklist.

This checklist is a beginning and is not all inclusive of electrical hazards in construction environments. It is intended to be used as an aid in analysis and discussion in applying PtD concepts during early stages of project design to help reduce construction workers' exposure to potential electrical hazards. It is the authors' hope that this will stimulate innovative thinking and further progress in identifying opportunities to apply PtD concepts to reduce construction workers' exposure to electrical hazards.

Item No.	Hazard/Exposure
1	Are there any existing overhead power lines or communications conductors within the perimeter of the construction area, or over access entrances to the construction area?
2	Are there any existing underground power lines or communications cables within the perimeter of the construction area?
3	Are there any existing underground power lines or communications cables within the perimeter of the construction area?
4	Are there any existing arc flash hazards in the construction area?
5	When will electrical systems become energized?
6	Will construction workers be exposed to uninsulated energized conductors during construction?
7	Other
8	Other
9	Other

efforts as we continue to explore how to better address electrical safety in construction environments.

Barriers in Construction

Some of the barriers that tend to impede achieving the objectives of PtD as applied to electrical safety; however, these impediments serve as opportunities for changing the electrical safety culture.

- The belief that safety is the job of a contractor and not the job of the design organization.
- The belief by design personnel that safety concepts add cost and delay schedules.

- Electrical design personnel may not be familiar with safety management principles, standards, and regulations.
- Design personnel have limited construction experience and/or knowledge.
- The belief that "qualified workers" do not make mistakes.
- Designers may not be familiar with workplace hazards and safe work practices.
- Design firms may be concerned that addressing worker safety in the design process will increase their liability.
- Facility owners may not understand the cost of retrofitting safe work practices and other control measures after the facility design is complete.
- Lack of codification or regulation for PtD.
- Lack of a Systems Safety Analysis application in construction work environments.

Each of these barriers needs to be addressed. All of them are real and impact the ability of concepts such as PtD to progress in an effective manner.

The Financial Burden

From a purely financial perspective, project owners and equipment manufacturers bear much of the financial costs associated with electrical injuries in construction. Construction company insurance costs are passed through to project owners. Electrical incidents that damage critical electrical equipment can delay project completion. Worker compensation laws limit employer liability when employees are injured on the job. Project owners and equipment manufacturers, on the other hand, are vulnerable to personal injury liability litigation from injured workers and their families. The further development and application of PtD concepts hold the promise of reducing not only construction worker injuries and their personal financial losses but also the costs of these injuries and their consequences borne by project owners and equipment manufacturers. By understanding where the hazards are within an electrical system, methods can be explored to reduce or eliminate the hazards before construction starts. The analysis process needs to include a roadmap for performing the analysis, what should be looked for, and at what stage of the design the analysis should be performed.

Countries and economies that have emphasized PtD in policies, education, and practice have demonstrated its effectiveness in reducing occupational fatalities. The impact on human lives is clear. However, when evaluating specific actions and investments, the effect may not be as clear, or due to the impression that probabilities are low, the value of the investments may not be as easily explained to those who control resources. However, the use of systematic processes and decision-making methods may help solidify the perception of the value of implementing a safety-conscious design. Simultaneously accounting for the positive impact on operational costs and system reliability of the same design decisions should further solidify the perception that the safer thing to do is the right thing to do.

Through individual and collaborative efforts, avenues to better address electrical safety via PtD can be explored. Some opportunities include the following:

- Working to ensure that electrical safety professionals participate in further refinements of the NIOSH PtD strategy and implementation program
- Influencing the education of engineers and designers in the concept of PtD, by adapting the Australian *Safe Design for Engineering Students* curriculum guide for both engineering students and experienced engineers
- Demonstrated support for ANSI/ASSP Standard 590.3 for guiding application of PtD
- Expanding Table 9.3 to be more inclusive of actions that could be taken to reduce electrical hazards in construction environments

There needs to be a shift in thinking of owners, construction management, and design organizations. The shift needs to be from compartmentalized thinking to thinking "Life Cycle Safety," through the full life cycle of an electrical installation, including:

- Construction
- Commissioning
- Operation
- Maintenance
- Renewal
- Decommissioning
- Demolition

Personnel who have the ability to review the design must be involved early enough in the project process to make an effective impact. They must be knowledgeable not only of the electrical safety concepts of how to work with electrical hazards but also of what could be done to reduce or eliminate the exposure to electrical hazards. This involvement needs to take place before drawings are completed or, in some cases, even before they are started. After drawings are completed and equipment purchased, the probability of making changes is virtually nil.

The extensive references in Further Reading below indicate that the concepts of PtD are mature; however, the application of these concepts to electrical safety has been limited. Designing the way we always have before, because it is easy, is no longer acceptable. Design processes for electrical equipment and systems impact electrical injury and fatality rates. In all work environments, to effect change in electrical injury and fatality rates, there has to be a change in design processes impacting electrical hazard exposures.

Crosstalk: Questions for Discussion

1 How do you establish design objectives to reduce exposure to hazardous electrical energy in capital projects?

2 How many incident investigations include recommendations for design changes to reduce exposure to hazardous electrical energy?

3 How can you educate management on the value of reducing exposure to hazardous electrical energy through engineering design solutions versus depending solely on safe work practices?

4 What are some tasks that currently depend on administrative controls that could be redesigned to reduce dependence on administrative controls to control exposure to hazardous electrical energy?

5 What are five examples of inherently safer electrical design options that your organization does not have in its facilities?

Further Reading

American Society of Safety Engineers (1994). *ASSE Position Paper on Designing for Safety*. Des Plaines, IL: American Society of Safety Engineers.

ANSI/ASSP Z10, *Occupational Safety and Health Management Systems*, American Society of Safety Professionals.

Bowen, J. (2001). *Introduction to Safety by Design*, 2001 IEEE IAS Electrical Safety Workshop, Toronto OT, February 2001.

Bugaris, R. (2017). Improving electrical safety in the workplace: applying prevention through design to voltage testing. *IEEE Industry Applications Magazine* 23 (3): 12–23.

CFR 29, OSHA 1910.119, *Process Safety Management of Highly Hazardous Chemicals, Explosives and Blasting Agents*, US Occupational Safety and Health Administration, Washington, DC ©.

Christensen, W.C. (2003). Safety through Design: Helping design engineers answer 10 key questions. *Professional Safety, Journal of American Society of Safety Engineers* 49 (3): 32–39.

Christensen, W.C. and Manuele, F.A. (1999). *Safety Through Design*. National Safety Council.

Construction (Design & Management) Regulations (2007). *Health and Safety Executive*, United Kingdom.

Council Directive 92/57/EEC Temporary or Mobile Worksites (1992). *European Union.*

Domitrovitch, T. (2020). Safety by design: transformer applications, *EC&M Magazine*, October 2020, pp. 44–48.

Ertas. *Prevention Through Design: Transdisciplinary Process*, Academy of Transdisciplinary Studies Texas Tech University Mechanical Engineering Department, Lubbock, Texas 79409: https://ciret-transdisciplinarity.org/ARTICLES/Ertas_fichiers/ptd.pdf.

Floyd, H.L. (2004), *Product Safety Opportunities in Reducing Occupational Electrical Injuries and Fatalities*, 2004 IEEE Product Safety Engineering Society Symposium, San Jose, CA, June 2004.

Floyd, H.L. (2008). Keeping young engineers grounded. *Consulting-Specifying Engineer* 32–34.

Floyd, H.L. (2010). The NIOSH Prevention through Design Initiative, *IEEE Industry Applications Magazine* (May/June 2010), pp. 14–16.

Floyd, H.L. (2011). From the Brink of Disaster: Lessons for Electrical Safety and Reliability, *NETAWorld Magazine*, Summer 2011, pp. 88–94.

Floyd, H.L. (2014). Leading-edge developments in advanced safety management and applications to electrical safety. *IEEE Transactions on Industry Applications* 51 (2) March/April 2015: 1303–1308.

Gambatese, J.A., Hinze, J., and Behm, M. (2005). *Investigation of the Viability of Designing for Safety*, The Center to Protect Workers' Rights, May 2005.

Hecker, S., Gambatese, J., and Weinstein, M. (2004). *Designing for Safety and Health in Construction.* Proceedings from a Research and Practice Symposium, University of Oregon Press.

IEEE, *Standard 1683-2014 – IEEE Guide for Motor Control Centers Rated up to and Including 600 V A.C. or 1000 V DC with Recommendations Intended to Help Reduce Electrical Hazards*, IEEE Standards Association, Piscataway, NY.

IEEE/ANSI C2, *National Electrical Safety Code*, Institute of Electrical and Electronics Engineers, Piscataway, NJ.

ISO 45001, *Occupational Health and Safety Management Systems*, International Standards Organization, Geneva, Switzerland.

Lee, B. (2014). *Industrial Intelligence – Reliability*, MARCON Conference, Knoxville, TN.

Lyon, B.K. and Popov, G. (2019). Risk treatment strategies – harmonizing the hierarchy of controls & inherently safer design. *Journal of Professional Safety* 64: 34–43. American Society of Safety Professionals.

Main, B.W. (2008). Social controls for reducing risks – observations on US and European approaches. *Professional Safety, Journal of the American Society of Safety Engineers* 53 (5): 41–49.

Manson, S., Hughes, R., Kirby, R., and Floyd, H.L. (2013). *Best Practices for Motor Control Center Protection and Control*, 2013 IEEE IAS Petroleum and Chemical Industry Technical Conference, Chicago, IL, Sept. 16–18, 2013.

Manuele, F. (2007). Prevention through Design: addressing occupational risks in the design & redesign processes, *By Design*, a technical publication of the engineering practice. Specialty of American Society of Safety Engineers (October 2007).

Manuele, F.A. (2013). Preventing serious injuries and fatalities, time for a sociotechnical model for an operational risk management system. *Journal of Professional Safety* 51–59.

Manuele, F. (2020). *Advanced Safety Management*, 3e. Hoboken, NJ: Wiley.

McClung, B., McClung, L.B., and Mohla, D. (2005). *Electrical Design – Refined for Safety*, Conference Record of the 2005 IEEE IAS Industrial and Commercial Power Systems Technical Conference, May 2005.

Melson, J.P. (2019). Electrical safety in developing countries. *IEEE Industry Applications Magazine*, Mar/April 2019, p. 6.

MIL-STD 882E Standard Practice for Systems Safety, U.S., Department of Defense, [Online]: http://www.system- safety.org/Documents/MIL-STD-882E.pdf.

Mohla, D., McClung, L.B., Rafferty, N.R. (1999). *Safety by Design*, Conference Record of the 1999 IEEE IAS Petroleum and Chemical Industry Committee Technical Conference, September 1999.

National OHS Strategy 2002–2012 (2006). Commonwealth of Australia, 2002. *Safe Design for Engineering Students*, Australian Government.

NFPA 70E, *Standard for Electrical Safety in the Workplace*, National Fire Protection Association, 1 Batterymarch Park, Quincy, MA 02169.

NFPA70, *National Electrical Code*, National Fire Protection Association, Quincy, MA.

NFPA70E, *Standard for Electrical Safety in the Workplace*, National Fire Protection Association, Quincy, MA.

OHSAS 18001, *Occupational Safety and Health Management Systems–Requirements*, British Standards Institution, London, UK.

OHSAS-18001, *Occupational Safety and Health Management Systems – Requirements*, International Standards Organization.

OSHA 3885, *Recommended Practices for Safety and Health Programs*, U.S. Occupational Safety and Health Administration, October 2016.

Payman, D. and Fotuhi-Firuzabad, M. (2012). *A Reliability-Oriented Outlook on the Critical Components of Power Distribution Systems*, 9th IET International Conference on Advances in Power System Control, Operation, and Management (APSCOM 2012), pp. 1–6.

Plan for the National PtD Initiative (2010). U.S. National Institute for Occupational Safety and Health, Washington DC, USA.

Roberts, D.A. (2019). Applying risk assessment at the worker level, *IEEE Industry Applications Magazine,* January/February 2019, pp. 18–24. ANSI/ASSE.

The link between productivity and health and safety at work, *Hesapro Background Research Paper*: http://www.hesapro.org/files/Background_Research.pdf.

Valdes, M.E., Bugaris, R.M., and Wellman, C.M. (2017). Behind the scenes of IEEE 1683: a look into the creation of a standard. *IEEE Industry Applications Magazine* 23 (1): 26–33.

Z590.3 (2019). *Prevention Through Design Guidelines for Addressing Occupational Hazards and Risks in Design and Redesign Processes.* Des Plaines, IL, USA: American Society of Safety Professionals.

10

The Intersection of Safety and Reliability

To achieve financial success, a company needs a number of supporting functions such as manufacturing, human resources, health, safety and environment (HSE), marketing, sales, and engineering. Each of these functions has management systems to enable performance aligned with the organizations' primary objectives. In the mid-1990s, industry consensus standards for HSE management systems were developed. Several standards development organizations published HSE management systems standards, based on continuous improvement quality management concepts championed by W. Edwards Deming.

The fundamentals of maintenance and reliability management have been known for years. There are numerous publications that speak to this subject. While the fundamental elements of maintenance management are similar to those associated with managing safety, the first industry standard to guide managing equipment maintenance and reliability was ISO 55000, Asset Management, published in 2014. The Society for Maintenance and Reliability Professionals (SMRP) and European Federation of National Maintenance Societies (EFNMS) have developed, harmonized, and published metrics for evaluating the health of maintenance and reliability management systems. The SMRP publication is titled Best Practices Metrics, First publication October 19, 2009.

There are industry standards and guidelines that discuss the importance of maintaining electrical equipment, development of maintenance strategies based on the business or operational consequences of failure, and inspections and predictive maintenance (PdM) techniques to identify precursors to failure. These standards include the following:

- NFPA 70B, *Standard for Electrical Equipment Maintenance*
- IEEE 3007.2, *IEEE Recommended Practice* for *the Maintenance of Industrial and Commercial Power Systems*
- ANSI/NETA MTS, *Standard for Maintenance Testing Specifications for Electrical Power Distribution Equipment and Systems*

Hidden Risk in Occupational Electrical Safety: Through the Lens of System Safety,
First Edition. H. Landis "Lanny" Floyd.

These standards emphasize that maintenance of electrical equipment and the safety of personnel who interact with this equipment are related – i.e., a quality maintenance program is essential to a robust electrical safety program. Collectively, they provide guidance on designing and implementing an electrical maintenance program. IEEE 902-1998 states, *To be successful, a preventive maintenance program shall have the backing of management* [which hints at the need to have a management process or system to enable sustainable performance. NFPA 70B emphasizes the need to identify critical equipment in Section 6.3: *Equipment (electric or otherwise) should be considered critical if its failure to operate normally and under complete control will cause a serious threat to people, property, or the product.* However, there is a critical concept missing – these two standards do not address the importance of integrating the electrical equipment maintenance program into a comprehensive reliability and maintenance management process. This concept is essential for the effective allocation of limited financial and human resources available for electrical equipment maintenance. This chapter addresses the opportunity for making better informed decisions when allocating limited maintenance resources. As we examine these strategies, we review the relationship between electrical safety and electrical maintenance, provide an overview of safety and reliability management systems, introduce methodology to help focus on critical maintenance needs, discuss maintenance considerations that can be addressed early in facility design, and discuss elements of effective implementation.

A New Era in Electrical Safety

While recognizing the need for electrical maintenance is not new, the context and consequences of electrical maintenance have changed, especially in how it relates to changing paradigms in electrical safety. A new era in occupational electrical safety emerged in the last decade of the 20th century. For the first 100 years of industrial electrification, the application of overcurrent protective devices was about the protection of conductors and circuit parts. For most of the 20th century, the primary objective for the maintenance of overcurrent protective devices was to help ensure protection for electrical conductors and equipment. The safe work practices, administrative controls, and application of PPE provided in NFPA 70E *Standard for Electrical Safety in the Workplace* assume the equipment has been installed, commissioned, and maintained to ensure designed functionality. In recent years, the increasing knowledge and understanding of the arc flash hazard has significantly changed the relationship between electrical safety and electrical maintenance. The arc flash hazard analysis and the selection of thermal PPE are completely dependent on the protective devices to function exactly as designed. In other words, the maintenance of certain devices and systems is critical to the safety of personnel who interact with the electric power equipment. CBs must function as designed. Overcurrent devices must operate at the designed and documented pick-up and time settings. If CBs or protective devices are dependent on an external power supply, the tripping power system (usually batteries and a battery charger) must be functioning as designed. If the protective device is a fuse, the installed fuse must meet the design specifications and be the type, class, and rating of the one documented in the arc flash analysis. There should be a robust management system to ensure no deviation from this expectation. In addition, the impact of failed electric power equipment needs to be broadened to include safety and environmental incidents that may occur in the manufacturing process. This is one connection point between the people and systems that focus on the manufacturing process and equipment and those that focus on electric power equipment.

Maintenance and Reliability Management Systems

A discussion of what makes an asset reliable can take many different directions depending on the group having the discussion and whether the discussion is about manufacturing high hazardous materials, discrete parts, maintaining infrastructures and buildings, or asset management.

Electrical power equipment typically does not have a lot of redundancy, especially as you get closer to the electricity consumer like a motor. Also historically, electric power equipment has a high reliability, which tends to result in an out of sight out of mind mentality for many in technical, operations, and even maintenance personnel. Here is where the electrical safety management process comes together with maintenance and reliability management processes. For electrical power equipment, if the frequency of predictive and preventive maintenance (PPM) tasks is not fully understood and integrated into the maintenance work management process, then these tasks can be forgotten or delayed beyond the useful life, and unexpected failure will occur. The unexpected failure can be apparent, such as an arc fault due to a loose connection, or it can be hidden, such as a power CB with an out-of-specification trip time. If the failure modes are not considered as part of a critical process system, evaluation of the driven equipment like pumps and compressors will be incomplete, thus opening a door for unexpected failure of the critical system.

Often, the engineers, managers, and technicians who focus on electrical equipment and electrical equipment safety are not the same as those who focus on process equipment reliability. The result is that it is not uncommon for either group not to have an appreciation for the concerns, interconnections, and processes of the other. This lack of appreciation can be resolved through basic cross-training and involvement in the various activities to develop the maintenance plans and reliability programs.

Maintenance Systems

The maintenance work first needs to be identified accurately. The identification steps can be separated into two parts. The first is the identification of the specific routine tasks that are performed during a PPM activity. The equipment may be operating or not operating while the routine tasks are being performed. An example of these activities would be inspection of contacts, thermographic evaluation, and replacement of a part(s) or oil analysis. A common and very effective process used in this step is the reliability-centered maintenance (RCM) analysis or failure modes and effect analysis (FMEA). During the analysis, a small group consisting of maintenance, operations, and subject matter experts are assembled in a team for the analysis. This is where an electrical power professional should be engaged to make sure the following questions are answered with consideration for the electrical power equipment:

- What are the functions and associated performance standards of the asset in its present operating context?
- In what way does it fail to fulfill its function?
- What causes each functional failure?
- What happened when each failure occurs?
- In what way does each failure matter?
- What should be done if a suitable proactive task cannot be found?

When considering that for an entire unit or plant, the job can become daunting, the value is often in question. However, it is important to remember the following:

- A hidden failure in an engineering system essential to control of hazardous electrical energy could have a serious, perhaps catastrophic, negative consequence in the effectiveness of established safe work practices or PPE.
- Asset downtime, whether scheduled or unscheduled, costs the business money and consumes resources. Unscheduled downtime is the most expensive and increases the potential for safety incidents. The analysis, assuming the recommendations are implemented properly, will reduce unscheduled downtime.

Whether the work is a required routine task like cleaning contacts, electrical equipment safety inspections, or a deteriorating condition picked up during a predictive inspection, accurate work identification during the work identification step will help facilitate a focused and efficient planning process. It will help in prioritizing the work planning sequence in the planning process. Also identifying the work as early as possible will provide time to consider all the safety aspects of each job. The safety aspect not only relates to the safety of the people doing the work, which is extremely important, but also allows time to consider additional safety and environmental implications that may occur upstream and downstream of the equipment being worked on. This is especially true when planning work on utility systems like electric power distribution. It is so interconnected with all aspects of the manufacturing process that a shutdown in one part of the system can have significant safety implications on another part of the plant.

Reliability – An Improvement Process

Defining the criticality of the equipment is important in setting priorities for preventive maintenance (PM) review, defining the PdM method, and developing reliability improvement programs. The next section identifies one aspect of criticality as it applies to personal exposure to shock or arc flash – a very important aspect in electrical safety. However, there are additional categories for criticality that need to be considered. Some of these categories are process safety management, environmental, maintainability and production rate, yield, and quality, as illustrated in Table 10.1.

Maintenance and reliability professionals have expertise in applying reliability improvement tools like data collection and analysis, RCM and FMEA, and reliability modeling. However, many also lack the basic understanding of the basics of electrical power equipment.

In summary, it is important for the people who manage the electrical power equipment to reach out to the people managing the manufacturing process

Table 10.1 Questions to assist identification and prioritization of electrical equipment maintenance.

Differentiating Electrical Maintenance
1. Is the reliability of this equipment or system essential for preventing or limiting personnel exposure to hazardous electrical energy? (Electrical safety critical)
2. Is the reliability of this equipment or system essential to assuring electrical energy or control systems for process safety or life safety systems? (Safety critical, possibly electrical safety critical)
3. Is the reliability of this equipment or system essential to assuring electrical energy or control systems for critical business operations? (Operations critical)
A "yes" answer to questions 1 and 2 identifies equipment or systems with high priority for maintenance, testing, and inspections to ensure the desired reliability necessary for personnel safety.

equipment to both educate them on electrical power equipment safety and maintenance. Conversely, it is important that the people who manage the maintenance and reliability improvement processes for the process equipment reach out to the electrical power equipment professionals to help them understand reliability management systems and tools.

Differentiating Equipment Critical to Personnel Safety

Two terms to help identify and differentiate electrical equipment that is critical to personnel safety are *electrical safety critical maintenance* and *electrical safety dependence maintenance*. These two terms can serve as tools to aid electrical safety and electrical maintenance personnel to identify critical equipment and maintenance activities to help achieve synergy and optimization of safety and maintenance objectives.

Electrical Safety Critical Maintenance

Electrical safety critical maintenance is directed to equipment, components, and systems whose failure directly impacts an electrical hazard control measure. Examples: fuses, CBs, certain protective relays, tripping power systems, GFCIs, and other equipment having the specific purpose of preventing or limiting exposure to hazardous electrical energy. The failure of this equipment results in the loss of an engineering control to protect people from electric shock and/or arc flash hazards. The key question to discuss relative to the is "Is the designed functionality of this equipment (or system) assumed in the protection of personnel from electric shock or arc flash hazards?" There are other reasons that the reliability of certain electrical equipment or systems may require a high priority relative to the broader population of installed electrical equipment.

Electrical Safety Dependent Maintenance

Electrical safety dependent maintenance is associated with the maintenance of equipment, components, and systems whose reliability and mean time between failures are key variables in the frequency of personnel exposure to electrical hazards. Example: motors, switchgear, lighting systems, and heating systems. The failure of this equipment results in personnel interaction with equipment having potential shock and/or arc flash hazards. For example, every motor failure initiates a series of events that entail maintenance worker exposure to electric shock and arc flash hazards. This may include operating disconnect switches, opening electrical cabinets to perform troubleshooting, performing diagnostic testing, disconnecting and reconnecting motor leads, and replacing fuses. Extending motor reliability reduces the frequency of these events, thus reducing the opportunity for serious electrical injuries. Effective application of PdM techniques, such as IR thermography, serves to reduce unscheduled emergency repairs, which tend to have a higher frequency of safety incidents than planned and scheduled work.

The terms *electrical safety critical maintenance* and *electrical safety dependent maintenance* do not currently appear in the mentioned industry standards for electrical safety and electrical maintenance. They can serve as tools to aid electrical safety and electrical maintenance personnel to identify critical equipment and maintenance activities to help achieve synergy and optimization of safety and maintenance objectives.

Inherently Safer Maintenance Technologies

Electrical maintenance technologies exist that can reduce personnel exposure to electrical hazards, identify deteriorated electrical equipment before failures occur, and provide remote alarms for faulted systems.

Remote Racking

Racking in and racking out of an electrical device can be defined as the act of connecting or disconnecting an electrical device from a main bus of the equipment by the intended racking means for that device. This task is routinely performed in medium-voltage switchgear and LV switchgear. Unless the work is performed during a maintenance outage, the equipment main bus is energized while this task is performed. Unless the switchgear is designed for closed door racking and is arc resistant per IEEE C37.20.7 or some other recognized arc-resistant equipment performance standard, the act of racking the device can expose the person performing this racking to arc flash hazards.

A remote racking device can be used to allow the person performing the racking operation to stand outside the arc flash hazard boundary. There are several types of remote racking devices. For each of these devices, a cord-connected control device is connected to the switchgear or racking device to enable operation of the racking device. Variations of remote racking devices include the following:

1) **Integrated:** These racking devices are an integral part of the CB or switchgear.
2) **Equipment specific:** These racking devices are designed to connect to a specific manufacturer and model of switchgear or equipment with clamps, bolts, or magnets.
3) **Universal:** These racking devices typically have a rotating shaft that can be connected to a CB racking mechanism. They have adapters that allow them to be used with a variety of CBs.

A remote racking device can be purchased either with new equipment or for existing equipment.

Remote Operation

Operating (i.e., opening and closing) an electrical device in energized equipment can expose the person performing that operation to an arc flash hazard. This task is routinely performed in medium-voltage switchgear, LV switchgear, and medium-voltage motor control centers. Unless the work is performed during a maintenance outage, the equipment main bus is energized while this task is performed. Unless the switchgear is arc resistant per IEEE C37.20.7 or some recognized arc-resistant equipment performance standard, the act of operating the device exposes the person performing this racking to arc flash hazards.

A remote operation device can be used to allow the person performing the device operation to stand outside the arc flash hazard boundary. Several types of remote operation devices are noted as follows:

1) **Remote control panel:** For devices that can be electrically operated (such as a substation CB or a motor-operated switch), a remote control panel can be installed outside the arc flash boundary. The remote panel typically consists of status lights to indicate the device position (such as "open" or "closed") and control switches for each device to be controlled. The control switches in the panel can be operated to open and close the switching devices while keeping the operator outside the arch flash boundary.
2) **Computer controlled:** For devices that can be electrically operated, a computer control system can be incorporated to allow device operation from a remote location. The control could be either through hard-wired connection for each device or performed using a communication network for the facility.
3) **Umbilical cord control:** For devices that can be electrically operated, an umbilical cord control can be used. The umbilical cord control scheme uses a pendant station with Open, Close, and Permissive controls, a 20′ or longer cord, and a plug-in connection to a specific CB control circuit. The umbilical cord control device is connected to the CB to be operated. The person operating the devices then stands outside the arc flash boundary and operates the CB using the pendant station controls.
4) **Portable control device:** For devices that are controlled by a pistol grip control switch or a pushbutton, a portable operating device can be temporarily secured to the equipment to allow remote operation of the switch or pushbutton. The portable device can have a 20′ or longer cord that connects the portable operating device to a control box. The control box is used to operate the portable operating device, which turns the switch or pushes the pushbutton to control

the equipment being operated. The person operating the controls can stand outside the arc flash boundary.

5) **Disconnect switch operator:** Disconnect switches are typically found in medium-voltage and LV motor control centers. Operating a disconnect switch can be done using a remote operator for that switch. The remote operator includes the operating device, a 20′ or longer cord, and a pendant station with the device controls. This allows the person operating the switch to stand outside the arc flash boundary.

Remote Monitoring

Organization may view substation inspections and maintenence as work that can be postponed or deferred as a cost control measure. This makes the detection of impending switchgear problems more difficult. Electrical equipment alarms can be incorporated into a remote monitoring system. That system can provide alarms to indicate when equipment problems occur.

One of the most effective means to provide alarms to a remote location is to use a communication network in the electrical equipment to provide access to those alarms. That significantly reduces the wiring to the remote monitoring system as compared to wiring each alarm to a remote location. Typically, only one communication cable is run from the monitored equipment to a remote communication network, such as a facility's Ethernet network. The remote monitoring system can poll all the equipment devices to check the status of all alarms. Metering and status information can be polled via this same communications cable.

Typical substation and medium-voltage motor control center alarms can include the following:

- Equipment status (open/closed, racked-in/racked-out)
- Protective relay status (healthy/unhealthy)
- CB trip coil status and control power status
- Lockout relay coil status and control power status
- Power quality events
- Ground fault alarm for high resistance grounded systems
- Battery system failures (loss of AC supply power, low DC voltage, high DC voltage, battery system ground)
- Impending equipment shutdown due to motor overload conditions
- Space heater circuit monitoring (alarm on low amperage or loss of control power)
- Transformer alarms (high temperature, high pressure, low oil)

Many of these alarms are signaled through dry contacts. These contacts can be wired to protective relay inputs or digital input devices. The relays and input devices can be connected to the equipment communication network, which allows the alarm status to be accessible through the network.

The remote monitoring can also be used to monitor system loads. This information can be used to:

- Monitor motor loads to identify low-load conditions (such as a failing pump)
- Automate load segregation systems based on local generation and critical loads
- Establish PM program based on equipment usage and performance (such as the number of start/stop operations)
- Protective relay status (healthy/unhealthy)
- CB trip coil status and control power availability

Smart Substations and Motor Control Centers

The integration of intelligent devices, communication networks, and diagnostic software into traditional motor control and power distribution substation equipment introduced a new era in how maintenance personnel interact with electrical equipment. These technologies can have benefits in installed cost, operational efficiencies, and safety enhancements to common operating and maintenance tasks. This technology enables many troubleshooting and maintenance tasks to be performed without exposure to energized parts and outside the arc flash boundary.

IR View Ports and Viewing Panes

IR cameras are routinely used to inspect electrical equipment for poor connections and overheating. The identification before failure of poor connections and overheated equipment allows repairs for that equipment to be planned and scheduled, leading to a safer work environment.

Since an IR camera cannot see through metal, often the only method available to access electrical connections in metal-enclosed equipment is to open compartment doors or remove compartment covers. Opening doors or covers to energized equipment usually exposes a person to electric shock and arc flash hazards. IR inspections of critical equipment are sometimes not performed due to the hazards of the task of accessing the equipment.

The use of an IR view port or viewing pane allows the IR inspection to be performed without the opening of the enclosure door or removing the enclosure cover. The view port or viewing pane is installed during a maintenance outage. The viewing pane provides a 2″–4″ viewing area that an IR camera can use to see inside the equipment. The view port provides a small opening that an IR camera equipped with a fish-eye lens can use to see inside the equipment. This allows the IR inspection to be performed without the need to open the equipment door or remove the equipment cover.

IR view ports and viewing panes can easily be installed in a variety of equipment, such as medium-voltage switchgear, LV switchgear, medium-voltage motor control centers, transformer terminal boxes, motor terminal boxes, and LV motor control centers. Depending on local inspection requirements, retrofitting these devices in existing installations may involve compliance with equipment testing and listing standards.

Insulating Oil Monitoring

Insulating oil is used in a variety of electrical equipment, including CBs and transformer. Oil-filled equipment should have oil testing performed at least annually to check for changes in the oil. Those changes can indicate equipment overheating, insulating problems, high levels of moisture, and internal arcing in the equipment. Testing consists of taking an oil sample and sending that sample to a certified laboratory for analysis. The laboratory provides a report of the condition of the oil and will indicate if any items need attention.

Continuous oil monitoring can be justified on critical equipment. Continuous oil monitoring systems are installed on the equipment and circulate oil through the device and continuously test certain oil parameters. An alarm can be provided if a certain parameter or combination of parameters is outside the defined limits. For some systems, a communication interface can be provided, which will allow for remote monitoring of all measured oil parameters.

Annual oil sampling will gradually reduce the oil level in the equipment. For liquid-filled transformers, an IR camera should be used to check the proper operation of transformer cooling fins. Each of the cooling fins should be approximately the same temperature. Cool fins typically indicate a low oil level, as there is not enough oil to enter the radiators. In some cases, only one radiator fin may be cool as compared to the others. That is enough to indicate that further oil-level checks should be performed.

Identifying and correcting low transformer oil levels will prevent unnecessary overheating of transformers (reducing transformer life) and unnecessary transformer failure (eliminating transformer replacement costs and production outage).

IR inspections can be used to get an approximation of oil level in CBs, but the results are not definitive due to the construction characteristics of the CBs.

Partial Discharge Analysis

Partial discharge analysis (PDA) can be used to monitor medium-voltage systems. The PDA system can detect small electrical discharges in the electrical system. An increase over time in the discharge rate and magnitude can indicate an impending failure. The monitored system can be shut down and inspected to identify the location of and perform repairs for the impending fault.

PDA systems can be used in medium-voltage equipment, such as switchgear, motor control centers, motors, and generators.

Temporarily Reduced Protection Settings

The first part of creating a safe work environment is to de-energize the electrical system. That work usually involves interaction with the energized switchgear. During that interaction, the reduction of protective relay and CB trip unit settings would usually provide faster clearing for faults in the equipment. Although selective coordination of the system may be compromised, that compromise may be acceptable during the interaction with the equipment.

Reducing the protective relay or trip unit settings can be accomplished in basically two ways:

1) **Manual switch:** A switch-on equipment can be used to select a different set of time current curves for a protective relay or CB trip unit. The switch could be either a separately mounted switch or one that is incorporated into the protection device.
2) **Automatic switch:** A motion or proximity sensor can be used to detect the presence of personnel in the area near the equipment. The sensor can select a different set of time current curves in the protective relay or trip unit, just like the local switch. A timer can be used to ensure the protection settings are switched to normal after a predefined amount of time from the operation of the motion or proximity sensor.

Online Temperature Monitoring

An online temperature monitoring system can be installed to monitor bus connections and other critical connections in equipment. Wireless temperature monitoring systems are available. The temperature monitoring systems provide continuous temperature information. Temperatures outside defined limits can be investigated.

Other Opportunities

The mentioned industry consensus standards relative to electrical safety and electrical maintenance could incorporate the concepts described in this chapter to provide aids in differentiating equipment and maintenance critical to personnel safety. One additional opportunity to enhance the maintainability of critical electrical equipment is to design the system with sufficient redundancy and flexibility to facilitate maintenance with minimal disruption to operational requirements. Accessibility and availability for inspection and repair are essential considerations when designing the electrical system. A given electrical system might include one or more of the following design features:

- Redundant equipment systems (transformers, switchgear, feeders) at the electric utility interface or at the main substation
- Primary selective and/or loop system design at the primary distribution voltage level to provide flexible operating configurations and maintenance isolation capability
- Dual unit substations with secondary selective (main-tie-main) designs to improve maintenance access
- Redundant charging systems for critical DC battery systems
- Alternate electrical power sources, such as a generator or uninterruptible power supply (UPS) system for critical loads (Maintenance priority should generally be given to existing alternate electrical power systems.)

As the facility design is being developed, these and other options should be considered, which can improve the overall maintainability of the electrical system and equipment. For existing facilities, these enhancements should be evaluated as part of the electrical system modifications and infrastructure upgrades.

Establishing the appropriate maintenance frequency of critical electrical equipment is essential to a successful maintenance strategy. An RCM analysis and an FMEA referenced previously can be used to determine how often equipment should be maintained. Maintenance frequency recommendations for specific electrical equipment are beyond the scope of this chapter. Helpful resources include Chapter 2 of NFPA 70E, NFPA 70B, and IEEE 3007.2.

Business and commerce are dependent on electrical technology for energy, control, data, and communications. In conducting this critical analysis of existing maintenance strategies using concepts described in this chapter, an organization can develop a better understanding to manage its electrical safety program more closely coupled with its maintenance and reliability program and will likely find opportunities to derive benefits across a broad set of business performance parameters that depend on defect-free operation of electrical energy, control, and communications systems critical to operations. Recent evolution in understanding the role engineered systems have on mitigating arc flash hazards is changing paradigms on the reliability of certain equipment critical for managing risks of both shock and arc flash hazards. This chapter introduced terms to help distinguish certain equipment and systems as having higher priority needs for maintenance to ensure reliability essential for electrical safety. We have discussed the opportunity to critique interaction and synergy linking electrical safety programs and maintenance and reliability management systems. The performance parameters that can benefit from an effective electrical safety program include improved energy utilization, improved on-time delivery, fewer environmental releases, optimum employee safety, improved raw material utilization, improved first-pass yield, and increased operations uptime. It is important to recognize the limitations of people expert in specific aspects of electrical safety, safety management systems, and maintenance management systems. By recognizing these limitations and engaging these experts in a collaborative process, organizations may discover hidden synergies and optimization opportunities.

Crosstalk: Questions for Discussion

1. How does your organization differentiate equipment maintenance impacting operations continuity from maintenance to ensure safety functionality?

2. How does your organization monitor the functionality of equipment critical to electrical safety?

3. What are opportunities to build synergy between safety and reliability objectives for your organization?

4. What are examples of nonelectrical equipment failures that trigger exposure to hazardous electrical energy?

5. How can you educate management on the effects of equipment reliability on your electrical safety program?

Further Reading

Allied Reliability Inc., *PdM Secrets Revealed,* 1st edition, March 28, 2006.

ANSI/ASSP Z10, *Occupational Safety and Health Management Systems,* American Society of Safety Professionals.

ANSI/NETA MTS, *Standard for Maintenance Testing Specifications for Electrical Power Distribution Equipment and Systems,* InterNational Electrical Testing Association, Portage, MI.

Cossé, R.E., Bowen, J.E., Combs, H.T. et al. (2005). Smart industrial substations. *IEEE Industry Applications Magazine* 11 (2): 12–20.

Doherty, M., Floyd, H.L., Graves, L.R., *Electrical Safety Management,* Conference record of Safety 2010, American Society of Safety Engineers, June 13–16, 2010, Baltimore, MD.

EEE 3007.2, *IEEE Recommended Practice for Maintenance of Industrial and Commercial Power Systems*, Institute of Electrical and Electronics Engineers Inc., Piscataway, NJ.

Floyd, H.L., *The NIOSH Prevention through Design Initiative, IEEE Industry Applications Magazine,* May/June 2010, pp. 14–16.

Floyd, H.L. (2012). Make the connection: relating electrical safety maintenance to reliability. *IEEE Industry Applications Magazine* 18 (6): 74–80.

IEEE C37.20.7, *IEEE Guide for Testing Metal-Enclosed Switchgear Rated Up to 38KV for Internal Arcing Faults,* Institute of Electrical and Electronics Engineers Inc., Piscataway, *NJ.*

ISO 14001, *Environmental Management System,* International Standards Organization.

NFPA70B, *Standard for Electrical Equipment Maintenance,* National Fire Protection Association, 1 Batterymarch Park, Quincy, MA.

NFPA70E, *Standard for Electrical Safety in the Workplace,* National Fire Protection Association, 1 Batterymarch Park, Quincy, MA.

Society for Maintenance & Reliability Professionals, *Best Practices Metrics*, First Publication, October 19, 2009, *8400 Westpark Drive, 2nd Floor*, McLean, Virginia.

Valdes, M., The Role of the Circuit Breaker, *Insights Magazine*, Publication of Independent Electrical Contractors Association, August 2009.

11

Residual Risk and the Psychology of Lower Order Controls

The hierarchy of controls discussed in Chapter 8 is ranked in effectiveness, with elimination considered the most effective, with effectiveness decreasing as you progress through the hierarchy. Engineering design choices, including elimination and substitution, are higher order controls, and warnings, administrative controls, and PPE are lower order controls. The deafferenting factor between the two categories is the level of dependence on human performance in implementing the controls. The higher order controls are applied in the design of the work environment and impact the risk of exposure to hazardous electrical energy for the life of the facility. Lower order controls must be applied for each specific task that has potential exposure. Proper application of lower order controls is dependent on actions of management and supervision to provide and enforce the controls and worker understanding of when and how to implement the controls. Keane quantified the level of effectiveness as noted in Table 11.1.

In the context of the hierarchy of risk controls, warnings, administrative controls, and PPE are sometimes referred to as lower order controls. In practice, the effectiveness of these controls can be overestimated due to a misunderstanding of the human factors that affect error likelihood in their implementation. This chapter explores one human factor, the limitation of attention, on residual risk inherent in lower order controls. The limitation of attention has implications for:

- identifying hazards,
- assessing risk,
- taking appropriate actions based on warnings,
- executing safe work practices, and
- proper selection and use of PPE.

Raising awareness and improved understanding of these topics can enable more effective risk assessment, quantification of residual risk, and more effective selection and combination of risk controls to achieve risk ALARP.

Hidden Risk in Occupational Electrical Safety: Through the Lens of System Safety,
First Edition. H. Landis "Lanny" Floyd.

Table 11.1 Relative effectiveness of controls in the hierarchy of controls.

Control	Effectiveness (%)	Comments
Elimination	100 (if the action is permanent)	If a hazard is permanently eliminated, there is no risk of exposure or injury for the life of the installation. No other controls are needed. Refer to the discussion of the hierarchy of controls in Chapter 8. Temporary controls such as de-energization and lockout tag-out to control exposure may be described as eliminating a hazard but may have residual risk that must be controlled with other means, so temporary controls are not 100% effective.
Substitution	Not quantified	The effectiveness of substation is completely dependent on the selected substation.
Engineering controls	70–90	The functionality of engineering controls, such as covers and enclosures, GFCIs, and fuses and circuit breakers, is typically dependent on an administration control, such as maintenance.
Administrative controls (including warnings)	10–50	To maintain effectiveness, administrative controls typically require significant resources over a long period of time.
Personal protective equipment	<10	PPE programs typically require multiple administrative controls, such as procedures and training, and are dependent on rigorous compliance by supervision and workers at risk of exposure.

The controls applied in design processes are sometimes referred to as higher order controls. They are applied in risk assessments conducted during the design and redesign phases of tools, devices, equipment, and facilities. The higher life cycle value and effectiveness of higher order controls are due to hazard elimination and substitution of less hazardous technology or engineering controls that take place in the design and redesign stages, which reduce the risk for the life of the tool, installation, or facility. The concept is that measures to reduce risk in the design and redesign of work environments are more effective than measures dependent on human behavior to recognize hazards and comply with safe work practices.

The higher order controls reduce dependence on controls that rely on real-time knowledge and action of employees in management and supervisory roles, as well

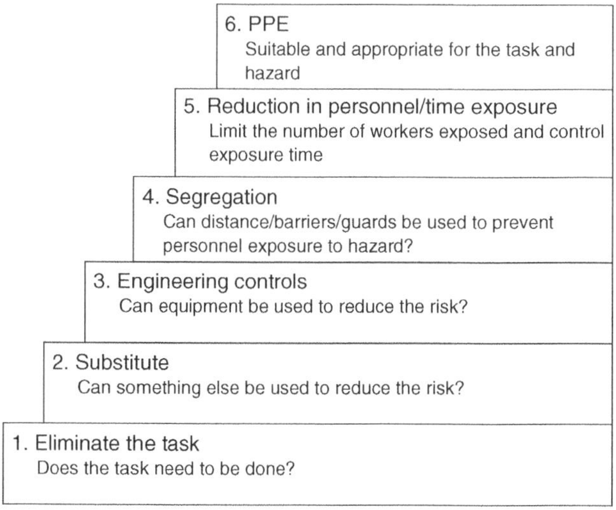

Figure 11.1 The hierarchy of controls applied to task risk assessment. Adapted from *Task Risk Assessment Guide*.

as the workers who are exposed to the hazards. The relatively lower effectiveness of the lower order controls (warnings, administrative controls, and PPE) is a function of the variability and reliability of people paying attention and responding appropriately when using warnings, administrative controls, and PPE. Selection of these lower order controls is typically made during risk assessments after design is complete to address residual risks not reduced ALARP in the design process.

Risk assessment after design processes are complete, and the work environment has been transformed from concepts to reality can take the form of task risk assessment. This application of risk assessment focuses on specific tasks, rather than the work environment. Figure 11.1 illustrates one approach, which shows step 1 as asking the question, "Can the task be eliminated?" Steps 2 and 3 mitigate the likelihood of mishap in ways that are independent of human error on the part of the worker at risk, like the control measures of substitution and engineering controls in Figure 11.1.

While lower order controls depend on workers' attention to recognizing hazards and doing the right thing at the right time to avoid injury, research from cognitive psychology (the field of psychology that studies attention) demonstrates that it is simply not possible for a human being to focus attention at all times and sustain attention for long periods of time. *These limitations of attention are not flaws in human behavior; it is a natural process of being human. To think otherwise is a misunderstanding of human nature that puts workers at risk.* The nature of human attention highlights some explanations of residual risk of lower order controls and, therefore, offers support for the importance of well-developed higher order controls in potentially hazardous work environments.

Power of Risk Assessment

The UK Health and Safety Executive defines risk assessment as the "need to think about what might cause harm to people and decide whether you are taking reasonable steps to prevent that harm." The rigor and discipline in the execution of risk assessment can vary widely and is likely a critical factor when comparing injury performance data.

Research by Mendeloff and Staetsky shows the occupational fatality rate in the United States is three times that of the United Kingdom. In the construction industry, their research shows the US fatality rate is four times that of the United Kingdom. For electrical contact or electrical discharge, the fatality rate is 3.8 times higher in the United States. The authors of the study note there may be cultural, reporting methods, and other factors that contribute to differences in safety performance. The study identified UK regulatory emphasis on risk assessment and rigorous application of a hierarchy of controls as a likely significant factor. This emphasis has spanned the full supply chain, assessing risk from tool, device and equipment design, facilities design, construction, operations, and maintenance.

The European Framework Directive on Safety and Health at Work (Directive 89/391 EEC), adopted by the EU in 1989, was a substantial milestone in improving workplace safety and health. The Directive introduced, as a key element, the principle of risk assessment with the priority of eliminating risk at the source and periodical reassessment of workplace hazards.

In the United States, the strategy and practice for improving workplace safety and health has taken a different approach. Emphasis has been more focused on compliance with product standards, building codes, safe work practices, regulations, and standards. This approach has placed greater dependence on lower order controls of warnings, administrative controls, and PPE than on the higher order controls.

Fred Manuele, a thought leader in safety management and a leading advocate for more effective application of risk assessment methodologies, stated, "Major and somewhat drastic innovations in the content and focus of occupational risk management systems will be necessary to achieve additional progress in serious injury and fatality prevention."

There is increasing awareness that the US occupational fatality rate is significantly higher than that for countries in which safety regulations and standards have placed a higher emphasis on risk assessment. Recognition of the power of risk assessment has been slower in the United States than in the EU and the

United Kingdom; however, there has been progress in the past decade. In 2005, C.A. Ericson published his book, *Hazard Analysis Techniques for System Safety,* in which he discusses in detail 22 proven techniques for identifying hazards and assessing risk. Mr. Ericson is recognized as a pioneer in risk assessment methodologies, having influenced advancements in risk assessment applications in aviation and space exploration, including the International Space Station. In 2007, the US NIOSH launched the PtD initiative with a strategy to bring greater attention to risk assessment, focusing on the application of higher order controls. In 2013, the American Society of Safety Engineers launched the Risk Assessment Institute and the Risk Assessment Certificate Program to develop risk assessment resources and competency to more effectively reduce the risk of occupational injuries and fatalities.

Specific to workplace electrical safety, the 2004 edition of NFPA 70E *Standard for Electrical Safety in the Workplace* added Annex F, Hazard/Risk Analysis Procedure. Subsequent editions expanded the original one-page Annex to eight pages, an indication of increasing recognition of the importance of risk assessments in electrical safety.

Widely considered one of the most prominent standards regarding workplace electrical safety in the United States, NFPA 70E, *Standard for Electrical Safety in the Workplace*, is limited in its scope to address "... electrical safety related work practices, safety related maintenance requirements, and other administrative controls." The effectiveness of these control measures is dependent on attention given to recognizing hazards and doing the right thing at the right time to avoid injury. If the residual risk of these control measures is not recognized and addressed, the effectiveness of these controls can be overestimated, thus exposing workers to unmitigated residual risks.

Risk Assessment and Hierarchy of Controls: An Application Example

During an electrical safety second-party audit of a chemical manufacturing facility in the US Midwest, the audit team discovered an opportunity to redesign an overhead 22-kV power line to eliminate the risk of contact during material handling activity. The 22-kV line was owned by the public utility. It was not on the plant property, but ran just outside the plant perimeter fence for approximately 300 feet. The vertical height was in compliance with the National Electrical Safety Code requirement of 14.5′ for terrain not accessible by traffic. The risk to plant employees and contractors was related to the area along the fence perimeter being used as a material and equipment storage yard. Movement of the material and equipment typically involved the use of fork trucks or small cranes. The plant managed the

Table 11.2 Residual risk after applying example control measures.

Control measure	Residual risk
Close Proximity Permit	• Likelihood that workers recognize the hazard • Likelihood that workers are trained in use of Close Proximity Permit • Likelihood workers understand how to use Close Proximity Permit • Likelihood Close Proximity Permit will be used 100% of time • Likelihood use of Close Proximity Permit is audited to measure understanding and compliance
Dismantle and eliminate the unused 22 kV line	None, the hazard was permanently eliminated

risk of overhead line contact with the use of a Close Proximity Permit. Discussion of this approach identified residual risks that did not have additional mitigation controls. Further discussion revealed that the 22-kV overhead line, although energized, was no longer serving any load. It was an extension from the public utility that had at one time served a pilot manufacturing facility that had been demolished more than 10 years earlier. Plant management contacted the electric utility and arranged to have the unnecessary 22-kV line removed. The impact of eliminating the unnecessary 22-kV line is compared to the residual risk of preventing contact using the Close Proximity Permit in Table 11.2.

Removal of the 22-kV transmission lines involved several hazards, some of which had significant levels of risk. However, the risk associated with the removal was assumed by the utility organization and their employees. This is a risk treatment option often referred to in risk management *as risk transfer, where the risk is shared with another party or parties.*

From a risk management perspective, such a transfer is acceptable for several reasons: (1) Once the hazard was removed, the associated risk was permanently eliminated and (2) the risk associated with permanently removing the 22-kV line was transferred to another party with more experience in managing risk relative to distribution lines. Just as the manufacturer's employees used lower order controls to manage the risk associated with energized distribution line contact, the utility organization's employees would use lower order controls during the removal of the transmission line. The reliability of lower order controls is related, in part, to the expertise of the people employing such means. It could be reasonably argued that utility workers are more practiced at employing lower order controls when erecting or dismantling distribution lines than manufacturers' material handlers are at avoiding contact with those lines.

In the electrical world in which we live today, it is rare to eliminate an electrical hazard completely. More often, the goal is to modify the design so that the likelihood of human errors, the severity of injury potential, and the need for PPE are at a practical minimum. Even though elimination may not be practical, discussion may lead to previously unrecognized opportunities to reduce risk.

Human Performance Limitations of Lower Order Controls

Research by Dekker and others has shown that it is inevitable that people will make mistakes in judgment and behavior that can lead to injury, even though they may be highly competent, qualified, and conscientious. Qualified workers – those deemed knowledgeable and skilled – do sometimes get injured in ways that their training and skill, if used properly, could have prevented. In light of this, after an electrical incident, it is not uncommon for supervisors to wonder why a qualified employee was not "paying attention."

Although it is not always the case that inattention or distraction played a causal role in an incident, it is possible. Further, in cases where inattention does play a causal role, it is not always the fault of the worker that they were not paying attention. Inattention and distraction are natural processes of human cognitive (psychological) function.

Human Performance: Attention as a Limitation of Lower Order Controls

Lower order controls (warnings, administrative controls, PPE), which require conscious effort and action on the part of administration, supervisors, and workers, necessitate that people *pay attention* for the control to be effective. For instance, adhering to warnings requires that a worker see a sign, consciously process the information on the sign, and determine that the sign is worth heeding. Administrative controls require that management take the time and economic costs for training and processes and that workers actively attend, listen, understand, and remember information. The use of PPE requires that management understands what equipment is needed, has provided that equipment, and supports its use among workers and that workers themselves understand when and how to use that PPE and are alert and attentive enough during their shifts to recognize when they should be using it.

In short, human performance plays a role in the effectiveness of the three lower order controls. One element of human performance – attention – plays a large role and is the focus of the human performance.

What Is Attention?

Attention most basically refers to a person's selecting and processing of information. This falls into two aspects: *focusing* (selecting what objects to attend to) and *concentrating* (holding that selection for an extended period of time). Imagine that you are driving down the highway. You focus your attention on the traffic patterns in front of you, to the sides, and behind you, as well as the lane markings and traffic signs on the shoulder of the road. You hold these within your attention until you reach your destination.

Now imagine that you have been driving for eight hours with not much of a break. It is midnight and you have not eaten or slept. Would this impact your ability to focus your attention on the road and hold that attention? What are some limitations in how well a person can select an object of focus and sustain attention?

Limited Resources: The Defining Feature of Attention

Cognitive psychologists are still in discussion over theories to explain exactly how people shift focus from one object to another and hold that focus over extended periods of time. But regardless of which theoretical perspective seems most valid, all of them are built on the foundation that our cognitive capacity for attention is limited due to a limited amount of brain resources. As a human being, you simply cannot pay attention to *everything* in your environment at once, all the time. This is why the brain has developed ways to (1) select what objects to attend to, that is, to focus and (2) hold attention on those objects, that is, to concentrate. For example, as you are driving down the road, a billboard advertisement might catch your eye. For a short period of time, you might look at the billboard, but quickly decide to bring your attention back to the road because the billboard is not relevant to driving safely down the highway. In this example, you have selected to attend to the road and to hold your attention on the road.

Sustained Attention

"Sustained attention" refers to concentrating on one thing for an extended period of time. In 1978, *The Lancet* published an article by Stuart and Rutherford, which documented the attention span of 1,353 medical students. According to their research, medical students' attention reached a maximum about 10–15 minutes into a class lecture and decreased from 15 minutes onward.

More recent studies look at the impact of sustained attention through measurement of reaction time (the amount of time it takes a person to react after a change occurs in their environment). Indeed, the longer a person tries to sustain attention on one thing, the slower they are to respond to changes in their environment. A review by Sarter et al. [1] shares that people working on tasks longer than

60 minutes slow their reaction time by several hundred milliseconds. Although this may seem like a small amount, a delay as small as 100 milliseconds translates to a stopping difference of 10 feet for a driver traveling 70 miles per hour.

Inattentional Blindness

One of the most interesting phenomena in all of cognitive psychology is *inattentional blindness*, which describes the experience of people looking at objects but failing to see them. You may be able to recall a time when you searched for your car keys for several minutes before realizing that they were "sitting right in front of you." On a more serious note, inattentional blindness also captures the experience of a driver at fault in an accident who exclaims that he "didn't see" an oncoming vehicle when pulling into an intersection, or expresses quite sincerely that a bicyclist just "came out of nowhere." For the skeptics among us who conclude from this that inattentional blindness is not a result of cognitive limitations but is a result of *not looking*, studies of inattentional blindness that use eye-tracking equipment confirm that it is quite possible for people to look directly at an object and not process what they are seeing. A related phenomenon is *cognitive tunneling*, which occurs when a person is so focused on one object that they lose awareness of the broader environment surrounding them.

Both inattentional blindness and cognitive tunneling have clear implications for workplace safety. Further, these cognitive phenomena are not limited to one-time events. A study by Castel et al. found that 76% of employees in one building at a large state university did not know the correct location of their nearest fire extinguisher and 74% did not know the correct location of the nearest fire alarm. Both objects were passed by employees on a daily basis, as employees in the study had worked in their building for an average of five years.

Both inattentional blindness and cognitive tunneling have clear implications for a variety of practices regarding safety in the workplace. Most importantly, as safety professionals, we need to recognize that inattentional blindness and cognitive tunneling are not flaws of human attention but are symptomatic of normally functioning cognitive systems. This brings us back to the need for well-designed and utilized higher order controls.

> The response to [the research on inattentional blindness] is to realize that these cases of failing-to-see are entirely normal.
>
> –Daniel Reisberg [2]

The path toward effective use of risk assessment and expanded use of higher order controls depends on understanding the limitations of human performance and residual risks associated with lower order controls.

Advancing Risk Assessment

There is increasing evidence that safety management systems built on a foundation of risk assessment and methodical mitigation of risk by comprehensive application of the hierarchy of controls is more effective than one built solely on compliance with safe work practices. If residual risk due to human factors impacting warnings, administrative controls, and PPE are not effectively addressed, risk reduction can be overestimated, resulting in overvaluation of reduction in injury frequency and/or severity. Momentum in efforts to advance the application of higher order controls and PtD should be accompanied by more diligent recognition and mitigation of residual risk inherent in lower order controls due to the dependence on unavoidable human error.

Factors of Attention

As we focus on designing safe workplaces, it is important to keep in mind the following limitations of attention:

- The limitations around attention in human beings do not signify flaws in human behavior but are characteristic of normal cognitive functioning.
- Reaction time slows down after long periods of sustained attention.
- It is possible for people to look directly at objects and not consciously see them.
- It is possible for people to be so focused on one object that they lose awareness of their larger surroundings.
- Understanding how human attention works can help us understand the residual risk around lower order controls and develop better policies and procedures around higher order controls.

Crosstalk: Questions for Discussion

1 What are the administrative controls needed to ensure an effective electrical PPE program?

2 What factors influence the effectiveness of an electrical maintenance program?

3 What can limit the effectiveness of a job safety analysis?

4 How can a recession or a downturn in business impact application of administrative controls?

5 What is required for a warning to be effective?

References

1 Reisberg, D. (2016). *Cognition: Exploring the Science of the Mind*, 6e. W.W. Norton & Co.

2 Sarter, M., Givens, B., and Bruno, J.P. (2001). The cognitive neuroscience of sustained attention: where top-down meets bottom-up. *Brain Research Reviews* 35 (2): 146–160.

Further Reading

ASSE. Risk Assessment Certificate Program, American Society of Safety Engineers, retrieved from the internet at http://www.asse.org/education/cra/ (accessed 15 August 2016).

ASSE Risk Assessment Institute. American Society of Safety Engineers, retrieved from the internet at http://www.oshrisk.org (accessed 15 August 2016).

Castel, A.D., Vendetti, M., and Holyoak, K.J. (2012). Fire drill: inattentional blindness and amnesia for the location of fire extinguishers. *Attention. Perception & Psychophysics* 74: 1391–1396.

Dekker, S. (2006). *The Field Guide to Understanding Human Error*. Ashgate Publishing Company.

Doherty, M., Floyd, H.L., and Graves, R. (2010). Electrical safety management. *2010 American Society of Safety Engineers Professional Development Conference* (13–16 June 2010), Baltimore, MA.

Drew, T., Võ, M.L.-H., and Wolfe, J.M. (2013). The invisible gorilla strikes again: sustained inattentional blindness in expert observers. *Psychological Science* 24: 1848–1853.

Ericson, C.A. (2005). *Hazard Analysis and Techniques for System Safety*. Wiley.

Eysenck, M.W. and Keane, M.T. (2010). *Cognitive Psychology: A Student's Handbook*, 6e. Psychology Press.

Floyd, A.H.L. and Floyd, H.L. (2017). Residual risk and the psychology of lower order controls. *IEEE Transactions on Industry Applications* 53 (6): 6009–6014.

Health and Safety Executive (UK). Controlling the risks in the workplace, retrieved from the internet at http://www.hse.gov.uk/risk/controlling-risks.htm (accessed 27 April 2017).

ISO 31000. Risk management – principles and guidelines, Section 2.25, Note 1, International Organization for Standardization, Geneva, Switzerland.

Keane, J. (2015). Preventing major losses. *Journal of Professional Safety* 60: 42–48.

Manuele, F. (2013). Preventing serious injuries & fatalities – time for a sociotechnical model for an operational risk management system. *Journal of Professional Safety* 58: 59.

Manuele, F. (2020). *Advanced Safety Management*, 3e. Wiley.

Medeloff, J. and Staetsky, L., p. 4.

Medeloff, J. and Staetsky, L., Table II, p. 7.

Medeloff, J. and Staetsky, L. (2014). Occupational fatality risks in the United States and the United Kingdom. *American Journal of Industrial Medicine* 57: 4–14.

NFPA 70E. Standard for Electrical Safety in the Workplace, National Fire Protection Association, Battery March, MA.

NIOSH. Prevention through design, retrieved from the internet at https://www.cdc.gov/niosh/topics/ptd/ (accessed 2 November 2016).

Stuart, J. and Rutherford, R.J.D. (1978). Medical student concentration during lectures. *The Lancet* 312 (8088): 514–516.

Task risk assessment guide, Step change in safety, Aberdeen, Scotland UK, 1999.

U.S. Occupational Safety and Health Administration. Recommended Practices for Safety and Health Programs, publication OSHA 3885, October 2016, retrieved from the internet at www.osha.gov/shpguidelines (accessed 1 November 2016).

12

Measure What You Need to Manage

How do you measure the effectiveness of your electrical safety program? Does the absence of injury equate to good safety? Every organization or business's function depends on metrics to enable better management decisions aimed toward achieving its goals and objectives. "What are the right metrics?" is an essential question for every profession and business function, and electrical safety is no different. Most safety professionals agree that metrics are essential components of effective safety management systems as they help evaluate, monitor, and control injury risk. Metrics help assess the effectiveness of risk controls, identify potential injury sources, support progress toward achieving goals, and track trends over time.

The first step in measuring the effectiveness of your program is to define indicators that will help track performance and progress. Next, employers, managers, supervisors, and workers need to establish and follow procedures to collect, analyze, and review performance data. The final steps include planning and implementing program improvements based on learning obtained.

Both *lagging* and *leading* indicators should be used. Lagging indicators generally track worker exposures and injuries that have already occurred. Leading indicators track how well various aspects of the program have been implemented and reflect steps taken to prevent injuries or illnesses before they occur.

Lagging and Leading Metrics

In terms of workplace safety, metrics are generally divided into leading indicators and lagging indicators. Lagging indicators, also known as outcome indicators, typically examine after-the-fact issues and include injury statistics employers must report to OSHA. On the other hand, leading indicators tend to be measures of efforts to identify and reduce risk. There are no regulatory requirements driving the use of leading indicators.

Hidden Risk in Occupational Electrical Safety: Through the Lens of System Safety,
First Edition. H. Landis "Lanny" Floyd.

Lagging indicators generally track injuries that have already occurred. Lagging indicators can include the following:

- Number and severity of electrical injuries
- Workers compensation claims, including claim counts, rates, and costs

Because electrical injuries are relatively rare compared to injuries from other hazards in the workplace, lagging indicators alone can be misleading in assessing the quality and effectiveness of your electrical safety program.

Leading indicators can include the following:

- Level of worker participation in program activities, such as reviewing and revising the electrical safety program
- Number of employee suggestions for improving the program
- Number of near-miss incidents reported
- Number and frequency of management walkthroughs of activity involving exposure to electrical hazards
- Number of workers, supervisors, and managers who have completed electrical safety training
- Completion of maintenance of equipment critical to electrical safety

Lagging and Leading metrics examine different types of data; each has advantages and disadvantages. For instance, because of consistent use across all industry, lagging indicators that examine outcomes are usually better for benchmarking. Leading indicators may lead to better intervention tools and can be indicators of future performance but are typically uniquely designed for the context in which they are applied. Thus, leading indicators may not be candidates for benchmark comparison.

Lagging and Leading metrics should be evaluated regularly, with results shared with management and workers, and their feedback used to improve performance. Look for opportunities to compare your organizations Lagging results with other divisions within your company and with business and trade associations.

TRIR is Misleading

Total recordable incident rate (TRIR) is a standard safety metric in the United States. It is defined as the number of work-related injuries per 100 full-time workers during a one-year period. TRIR is a lagging indicator as it is a measure of past performance. TRIR may not be an indicator of future performance. Consider the following scenario: An organization has not had an OSHA recordable injury from electrical hazards for 365 days. On day 366, a worker suffers an OSHA recordable electric shock injury. The one-year injury-free performance on day 365 changed overnight. Did the effectiveness of its electrical safety program change overnight? How useful is the safety statistics on day 365 in predicting the likelihood of injury on day 366? The answer is that past performance is not a predictor of future performance. Sports teams do not play by watching the scoreboard of past performance, and automobile drivers do not drive with their eyes focused on the rearview mirror. Knowing the first quarter's score is important, but that likely will not tell the team what the score will be in the remaining quarters. A driver with eyes fixed on the rearview mirror may have a clear view of the past journey, while a catastrophe may be looming ahead.

In the United States, LTIs from exposure to electrical hazards are low-frequency events. Less than 0.2% or less than 1 in 500 occupational injuries are from electric shock and arc flash exposures. Events that are low in frequency can mean that an organization does not have sufficient internal data on electrical injuries to be statistically meaningful. Consider two companies using TRIR as an indicator of the effectiveness of their respective electrical safety programs. In the past two years, one company has 100,000 employees and experienced five OSHA recordable injuries from exposure to electrical energy. The second company has 350 employees and has not had an OSHA recordable injury from electrical hazards in more than five years. Which company has the more effective electrical safety program? Which company is doing better in identifying and reducing risk? Looking at an individual worker in each of the two companies, which worker is more likely to have an electrical injury? The answer is not in the TRIR data. Due to the inherent low-frequency occurrence of electrical injuries, an individual or company may not recognize the potential for a fatal (high-consequence) injury and be blind to exposures and risk having high severity potential outcomes. The very low frequency of electrical incidents and the potential for an organization to be blind to the effectiveness of its electrical safety program underscore the need for a balanced scorecard of leading and lagging indicators for electrical safety.

An organization may have leading indicators at a high or broad level for its overall safety program. Still, due to the infrequency of electrical incidents, leading indicators for an overall safety program may not be sufficiently focused on factors critical to electrical hazard exposures. No matter where an organization is in its electrical safety management journey, there are opportunities for a more balanced scorecard of electrical safety indicators.

Challenging the Norm

For more than 50 years, thought leaders in safety management have raised concerns about the limitations of depending only on lagging indicators for safety management. Peterson examined studies from the 1970s in the United States,

Great Britain, and South Africa that showed a weak correlation between past performance and future performance. Kaplan and Norton published a widely acclaimed book on the topic in 1996. More recently, contemporary thought leaders in advanced safety metrics, including Manuele, Lopez, Esposito, and Walaski, have brought heightened attention to a more balanced approach in designing and applying leading and lagging indicators.

The search for better methods and data to manage safety programs has accelerated significantly in recent years. Documentation of the examples below is included in the Further Reading for this chapter.

Work Safe Alberta

In 2020, the provincial government of Alberta, Canada, updated its publication *Leading Indicators for Workplace Health and Safety*. This document is a user's guide of what leading indicators are and how and why they might be applied to strengthen your organization's health and safety culture and performance. Some leading indicators may be more effective than others, depending on an organization's safety culture and maturity. For that reason, the user guide considers three categories of performance: those that are focused on maintaining compliance with legislation, those looking for improvement (beyond compliance), and those in pursuit of continuous learning for the highest levels of performance. A tool is provided to help establish which category best describes an organization's safety culture and performance.

American Society of Safety Professionals

The ASSP has highlighted advancements in metrics to manage safety programs better. The ASSP provided leadership and support to the 2022 revision of *ANSI Z16.1, Method of Recording and Measuring Work Injury Experience*, which provides guidance in designing a balanced approach using leading and lagging indicators.

National Safety Council

Drawing from the experience of high-performing member companies of the NSC, the Campbell Institute of the NSC published a collection of research papers to help organizations effectively put leading indicators into practice to protect workers. The design and implementation guides provide a comprehensive listing of leading indicators categorized by the maturity of an organization's program and the complexity of collecting and analyzing data. Included are sample leading indicators for employers in the categories of education/awareness, reach, participation, satisfaction, and organizational health.

US Occupational Safety and Health Administration

In 2019, OSHA created a website with resources aimed at helping employers use leading indicators, along with traditional lagging indicators, to track workplace conditions and events to prevent injuries before they occur and help employers improve their safety programs. The resources include a downloadable document with information on the characteristics of useful leading indicators, finding leading indicators in data that may already be collected, and how to use leading indicators to improve parts of a safety management program, as well as an action plan checklist.

Center for Chemical Process Safety

The Center for Chemical Process Safety (CCPS) is a technology alliance sponsored by the American Institute of Chemical Engineers. In 2019, the CCPS updated its guide for designing and implementing leading and lagging indicators for managing the risk of hazardous chemicals. The guide differs from the abovementioned references in that its focus is on a specific hazard. In that regard, it provides an example of applying generic guidance, found in the abovementioned references, to a unique hazard.

A common theme linking the abovementioned resources emphasizes that a combination of lagging and leading indicators can provide a balanced scorecard for managing continual improvement in occupational safety management. A balanced scorecard aims to provide insight into an organization's electrical safety initiatives. A balanced scorecard involves combining leading indicators with lagging indicators to develop a more comprehensive assessment of electrical safety performance and how it can be improved. Lopez, Esposito, Walaski, and others are helping educate safety professionals in the design and application of balanced scorecards for safety management. Standard *ANSI Z16.1, Method of Recording and Measuring Work Injury Experience*, originated in the 1930s. The latest revision reflects the evolution and recent research impacting lagging and leading indicators and the design and implementation of a balanced safety metrics scorecard.

For more than 10 years, leading safety organizations, including the NSC, the ASSP, and thought leaders in safety management, have brought attention to workplace hazards that have an exceptionally high potential of disabling or fatal injury. The intervention strategy, Serious Injury and Fatality (SIF) prevention, is designed to help employers ensure that low-frequency, high-consequence risks that comprise a small component of TRIR risks are appropriately managed. An organization will typically identify 8–10 hazards in its operations with SIF potential. The dangers with SIF potential may vary depending on the nature of operations, but exposure to hazardous electrical energy is common to nearly all industries and processes. SIF prevention programs help ensure top management is fully committed to the effective management of low-frequency, high-consequence hazards. Having top management support for an SIF prevention program will help ensure support for a more effective electrical safety program.

Hard Risk and Soft Risk

Risk can be a "hard" risk, such as the risk of electric shock when working near an energized circuit. Risk can also be a "soft" risk, such as the risk of having an ineffective incident investigation process. Hard risks tend to have defining factors and statistical data that closely connect a hazard to an undesirable outcome. For a person standing within inches of an unguarded energized conductor, the risk of suffering an electric shock is a hard risk. There is a clear and present danger.

Soft risks are created by actions and decisions that are distant in time from an exposure to a hard risk. The danger to a worker may not be clear and present. Soft risks are much harder to make decisions about as they tend to involve situations where the risk factors are unknown or otherwise difficult to determine. Thus, the course of action is difficult to decide. These generally take more planning and application of extensive risk management applications to weigh what is known against the majority that is not. A manager's lack of support to approve a budget to incorporate inherently safer designs in a capital project for a new installation is a soft risk. The manager's action will impact the effectiveness in managing the risk of electrical injury for the installation's life, but the likelihood of injury may be difficult to quantify.

Leading indicators tend to be measures of efforts to identify and reduce risk. Since hard risks may be more visible and easier to quantify than soft risks, identifying and understanding the relevance of soft risk may provide significant opportunities for identifying meaningful leading indicators.

Electrical safety leaders should educate themselves on the evolving body of knowledge in leading indicators and help coach management leadership to establish a balanced scorecard of metrics that can more effectively serve the organization in identifying and controlling the risk from electrical hazards. In particular, leading indicators should encompass all workers exposed to electrical hazards and low-frequency/high-consequence exposures as discussed by Gammon, Brenner et al. Specific attention should be given to soft risks to identify useful leading indicators. While TRIR may be a metric commonly used across most organizations, leading indicators may need to be different for different organizations. Additional examples of leading indicators of electrical safety program effectiveness include the following:

1) The existence and vitality of electrical safety leadership at the corporate; business unit; global region, and local; and organizational unit as appropriate
2) Quality and effectiveness of electrical safety training for electrical workers, nonelectrical workers, line supervision, and critical supporting resources; quality and effectiveness of training are more challenging to measure than the frequency of when training was provided.
3) The number of managers and supervisors trained in electrical hazards awareness and the electrical safety program's role
4) Frequency of program audits that examine the implementation of the organization's electrical safety program
5) Quality and frequency of management system audits focused on preventing exposure to electrical hazards
6) The percentage of capital projects giving attention to inherently safer design in the selection of hardware and electrical system design
7) Discipline in maintaining maintenance programs for equipment and systems critical to electrical safety, especially during business downturns when maintenance budgets may be reduced
8) The percentage of incident investigations that include prevention through design solutions, as opposed to more training, more procedures, and more PPE
9) The frequency of management and worker engagement in a program targeting hazards with high potential for SIF

The abovementioned list provides a few examples when designing potential leading indicators. The Work Safe Alberta publication *Leading Indicators for Workplace Health and Safety* and the OSHA publication *Using Leading Indicators to Improve Safety and Health Outcomes*, provide easy-to-use guidance to help an organization expand its use of leading indicators. As an example of their direction, Section 2.3 of the OSHA document guides the definition of leading indicators for each of the following elements of safety management:

- Management leadership
- Worker participation
- Hazard identification and assessment
- Hazard prevention and control
- Education and training
- Program evaluation and improvement
- Communication and coordination for host employers, contractors, and staffing agencies

Leading indicators should be tightly connected to electrical safety. For example, an organization with a leading indicator measuring management/leadership for overall safety and health should develop a metric specifically measuring management/leadership commitment to electrical safety.

The relatively low frequency of electrical incidents can blind an organization to risk reduction opportunities. Lagging indicators tell us where we have been, but not where we are going. Safety management experts and organizations championing new thinking in safety management are pointing to innovations in designing better lagging and leading indicators to create a balanced scorecard of safety metrics. An organization may have leading indicators at a high or broad level for its overall safety program. Still, due to the infrequency of electrical incidents, leading indicators for an overall safety program may not be sufficiently focused on factors critical to electrical hazard exposures. No matter where an organization is in its electrical safety management journey, there are opportunities for a more balanced scorecard of electrical safety indicators.

Electrical safety leaders should embrace advancements in safety metrics and develop a balanced scorecard tailored to their unique work environment and management culture, incorporating lagging and leading indicators to help better manage continual improvement in identifying and reducing hard and soft risks associated with exposure to hazardous electrical energy

An organization intent on controlling risk to ALARP should study the recent work exploring the leading indicators listed as references. Work Safe Alberta, OSHA, and the Campbell Institute have provided practical guidance to help an organization design and implement meaningful metrics for reducing hard and soft risks. A balanced scorecard of lagging and leading indicators can enable measurable and long-lasting improvements to electrical safety outcomes in the workplace.

Crosstalk: Questions for Discussion

1. What metrics does your organization currently use to monitor electrical safety performance?

2. What leading indicator metrics does your organization currently use for overall safety performance?

3. What processes are used to review electrical safety metrics with management and workers?

4. How many electrical incidents with injury has your organization experienced in the past five years?

5 How many near-miss incidents have been reported, reviewed, and program improvements identified in the past five years? How do you feel about the answer?

Further Reading

US Bureau of Labor Statistics Injury (2020). Illness, and Fatality Report retrieved from the internet at https://www.bls.gov/iif/ (accessed 12 December 2020).

Peterson, D. (2001). The safety scorecard: using multiple measures to judge safety systems effectiveness. *EHS Today* (30 April 2001). Retrieved from https://www.ehstoday.com/safety/article/21909343/the-safety-scorecard-using-multiple-measures-to-judge-safetyrnsystem-effectiveness.

Kaplan, R.S. and Norton, D.P. (1996). *The Balanced Scorecard: Translating Strategy Into Action*. Harvard Business School Press.

Manuele, F.A. (2009). *Leading & Lagging Indicators: Do They Add Value to the Practice of Safety?* Journal of Professional Safety, 28–33. Park Ridge, IL: The American Society of Safety Professionals.

Gary Lopez, C. (2020). *"Leading and Lagging Indicators," Journal of Professional Safety*. American Society of Safety Professionals.

Esposito, P.A. (2018). *Safety Metrics: Corporate & Site-Level Scorecards, Journal of Professional Safety*. Park Ridge, IL: American Society of Safety Professionals.

Walaski, P. (2020). "*The Role of Leading and Lagging Indicators in OSH Performance Management*," *Journal of Professional Safety*. American Society of Safety Professionals.

Work Safe Alberta (2020). *Leading Indicators for Workplace Health and Safety*. Canada: Government of Alberta.

ANSI Z16.1 *Method of Recording and Measuring Work Injury Experience*. Park Ridge, IL: American Society of Safety Professionals.

The Campbell Institute (2019). *An Implementation Guide to Leading Indicators*. Chicago, IL: National Safety Council.

The Campbell Institute (2019). *Beyond Safety: Leading Indicators for Health and Well Being*. Chicago, IL: National Safety Council.

The Campbell Institute (2019). *Elevating EHS Leading Indicators: From Defining to Designing*. Chicago, IL: National Safety Council.

The Campbell Institute (2015). *Practical Guide to Leading Indicators: Metrics, Case Studies, and Strategies*. Chicago, IL: National Safety Council.

The Campbell Institute (2013). *Transforming EHS Performance Through Leading Indicators*. Chicago, IL: National Safety Council.

Occupational Safety and Health Administration (2019). *Using Leading Indicators to Improve Safety and Health Outcomes*. U.S. Occupational Safety and Health Administration.

Center for Chemical Process Safety (2019). *Process Safety Metrics: Guide for Selecting Leading and Lagging Indicators, v3.2*. New York: American Institute of Chemical Engineers.

Gammon, T., Vigstol, D., and Campbell, R. (2019). Workers at risk of fatal and nonfatal electrical injuries. *IEEE Transactions on Industry Applications* 55 (6): 6593–6602.

Brenner, B., Cawley, J.C., and Majano, D. (2020). Electrically hazardous jobs in the U.S. *IEEE Transactions on Industry Applications* 56 (3): 2190–2195.

Floyd, H.L. (2022). Considerations for a balanced scorecard of leading and lagging indicators for your electrical safety program. *IEEE Industry Applications Magazine* 28 (3): 16–20.

Floyd, H.L. and Cole, B.C. (2015). Twenty years later: creating a continuous improvement environment for electrical safety. *IEEE Transactions on Industry Applications* 51 (3): 2633–2640.

13

End Game

> In my 40+ career with a large global company, the first 20 years included 12 electrical fatalities. The last 20 had zero. The cultural shift that achieved this improvement was inspired by the vision that eliminating electrical injuries was possible, and was led by passionate individuals who were influenced first hand by the tragedy of the fatalities and serious injuries. In parallel with this improvement, occupational electrical injuries in the US have steadily declined for the past 40 years, with more than 70% reduction in electrocution fatalities. With the improved performance, the tragic reminders become less frequent. How do we maintain discipline and rigor so that electrical safety leaders, individual companies and organizations do not drift from their vision?
>
> –Lanny Floyd

> When you … think you have this electrical safety problem licked, you will be in your most vulnerable position. You will start to let your guard down. Electricity is very patient and will wait a long, long time for people to let their guards down. And it will strike swiftly with no warning. You must never think you have finished your work on electrical safety.
>
> –Thomas D. Sterner

Tom Sterner, Director of Safety, Delmarva Power and Light Co., made this statement at an electrical safety conference held in 1988 in Wilmington, DE. His comments foreshadowed a future that would continue to see declines in the number of electrical injuries but could create an inaccurate perception of risk. The success we have seen in reducing serious electrical injuries makes his statement all the more ominous 25 years later. Certainly, electrical safety programs must be impacted when people involved with them have never seen a serious injury or fatality firsthand. Could the absence of serious injuries contribute to a *cultural drift* with a subsequent *occlusion* of electrical safety?

Hidden Risk in Occupational Electrical Safety: Through the Lens of System Safety,
First Edition. H. Landis "Lanny" Floyd.

Cultural Drift

"...the basic values, norms, beliefs and practices that characterize the function of an institution. At the most basic level, organizational culture defines the assumptions that employees make as they carry out their work. It is a powerful force that can persist through reorganizations and the change of key personnel. It can be a positive or negative force." – Columbia Accident Investigation Board [1]

Cultural Drift

"The gradual return to old ways of doing things, especially after a period of growth or progress. It's a slow slide into the habits that once held the organization back. You may not notice it at first, but over time, the effects include less innovation, less engagement, and fewer people willing to speak up" .[2]

Occlude

to close up or block off – Merriam-Webster Dictionary

How can we maintain an *organizational culture* around electrical safety built on motivation, passion, and understanding of vulnerability without witness to serious injury or fatality? This chapter discusses these questions and proposes solutions.

In 1989, the top management of DuPont, a global science and technology company, made a highly visible commitment to reduce the risk of injuries to employees and contractors from electrical hazards. Goals for sustainable improvement were established, financial support was provided, and dedicated people were empowered to reduce the probability of electrical incidents, injuries, and fatalities. One part of the effort was to change the culture to embrace continuous improvement.

In 1990 and 1992, several leaders in the company's electrical safety improvement initiative collaborated on two award-winning papers presented at the annual IEEE IAS Petroleum and Chemical Industry Conference and subsequently published *in IEEE Transactions on Industry Applications*. The first paper, *Maintaining Safe Electrical Work Practices in a Competitive Environment,* was presented at the 1990 IEEE IAS Petroleum and Chemical Industry Conference in Houston, TX. This paper described management recognition of deteriorating electrical safety performance and the creation of an organizational infrastructure to enable and support changes to better manage electrical hazards in company facilities and operations.

The second paper, *Creating a Continuous Improvement Environment for Electrical Safety*, was presented at the 1992 conference in San Antonio, TX. This paper outlined a strategy for establishing a culture for long-term continuous improvement in electrical safety. The elements of that strategy, shown in Box 13.1, describe an organizational culture intended to have a positive impact on preventing electrical incidents and injuries and are discussed in detail in follow-up papers by two of the original authors.

Box 13.1 Elements of the Strategy Described in the Paper, *Creating a Continuous Improvement Environment for Electrical Safety*

- Understand the business consequences of electrical incidents
- Engage all employees
- Stimulate near-miss reporting
- Apply quality improvement model – Plan-Do-Check-Act
- Build networks
- Challenge accepted practices
- Improve collaboration among management, electrical experts, and safety professionals
- Use standards as tools
- Promote prevention by design
- Address life cycle: design, construct, operate, maintain, dismantle

The culture and continuous improvement strategy described in these papers and nurtured for more than 20 years has resulted in significant improvement in the severity and frequency of electrical injuries in the company. Most dramatic is the impact on the frequency of fatalities from electrical energy. Prior to 1993, fatalities from electrical energy were occurring on average every 33 months. From 1993 to 2016, there were zero fatalities from exposure to hazardous electrical energy in the company operations.

Central to the strategy was the development of a company standard for electrical safety that integrated industry standards and best practices from around the world with the company's proprietary OSHMS.

Occupational electrocution fatalities as a whole in the United States have shown a steady decline since 1980. As the performance gets closer to zero, we need to consider the possibility that lower frequency of fatal injuries could contribute to a cultural drift away from safety vigilance.

Were the words from Sterner quoted in the opening of this chapter a forewarning to be alert for cultural drift? At that time, serious electrical injuries and fatalities in the example company's operations were real and tangible. The individuals leading efforts in electrical safety improvement had personal experiences involving serious injuries and fatalities to people they knew. This impacted their motivation, passion, and understanding of vulnerability. They worked hard to improve electrical safety programs. And they were successful. Today, in US workplaces, there are fewer serious electrical injuries and fatalities than ever before. But with this improvement, and a newer generation of people working on electrical safety programs, we must pose the question of how losing those real and tangible experiences impacts our understanding of risk.

High-Profile Example of Cultural Drift

In 2005, the National Aeronautics and Space Administration (NASA) published a report outlining the problems within NASA that led to the Columbia space shuttle disaster. In the report, they referenced the Challenger space shuttle disaster from 17 years earlier:

> The changes and protections instituted at NASA after the Challenger accident slowly degraded to the point where the same performance pressures and unrealistic expectations implicated in the Challenger accident contributed also to the Columbia loss.

Also in 2005, the International Organization for Economic Cooperation and Development studied several major industrial accidents and noted a common factor of *cultural drift*, the "systematic organizational performance deteriorating under competitive pressure, resulting in operation outside the design envelope where preconditions for safe operation are being systematically violated." Based on these reports, we can see that cultural drift can lead to these large-scale disasters that have been studied extensively to determine causal factors. Cultural drift as a contributing factor to accidents is not, however, limited to large-scale disasters. Individual accidents, including both nonfatal and fatal, may also be, in part, attributed to deterioration of systemic organizational performance.

Underlying Cultural Drift

The definition cited for cultural drift identifies the impact of competitive pressures. Are there other "pressures" or forces that could impact cultural drift specific to electrical safety? If we could identify these forces, could we mitigate and reverse their impact?

One factor that may contribute to cultural drift regarding electrical safety is a general misperception of risks associated with electrical hazards.

The relative infrequency of electrical injuries can create an illusion of having an effective electrical safety program, potentially creating *an occlusion of electrical safety*. A rare serious electrical injury or fatality may be incorrectly viewed as a random event, as illustrated in the quote in Box 13.2.

Box 13.2 Example of a Manager Who Had an Inaccurate Understanding of the Risk and Consequence Severity of Electrical Injuries

How could this have happened? We have the best safety record in our division. We've never had a lost time electrical injury at this plant. He is the most knowledgeable person in the crew. He has never had a recordable injury. I was not aware that my employees were exposed to a hazard with such severity. I don't understand. –A plant manager's comments during an arc flash injury investigation

Psychology of Risk in Cultural Drift

Because the frequency of electrical incidents is low, and the consequences of electrical injuries are typically severe, one of the biggest challenges within electrical safety is dealing with the need for workers and management to stay vigilant about this potentially severe hazard that occurs rarely. As described by the scientists in the space shuttle Columbia report, and echoed in the opening quote of this paper, when a significant amount of time goes by after an accident, it is human nature to relax, lose vigilance, and repeat the same actions that led to the initial accident. Why?

To understand how people can drift away from a safety-vigilant culture, it is important to understand how individuals process risk. This will help us explain how a person can relax their risk perceptions over time and can help us understand how to reverse or prevent this drift.

Our Mental Calculators: Two Methods of Processing Risk

When we judge our risk of experiencing an event, including an adverse event like an electrical injury, there are two mental systems we use. One is an *analytic system* (which is logic oriented, governed by conscious thought processes, and takes considerable time to reach a decision due to conscious weighing of all variables). The other is an *experiential system* (which is "affective" or feeling based, governed by associative connections and "vibes" from previous experiences, and takes very little time to reach a decision).

Whether we rely primarily on the analytic system or the experiential system varies from person to person and situation to situation, but almost always, no matter how logical we think ourselves to be, the experiential system prevails. This is particularly the case when the decision-making is complicated, when we have little time to make a decision, or when our mental resources are taxed. Whenever we are tired, hungry, or otherwise mentally preoccupied (that is, almost all the time), we will use the experiential system.

The Availability Heuristic: Making Decisions Based on What Comes to Mind

Our perception of risk of a particular event is somewhat based on our perception of that event's frequency. But how do individuals form a perception of an event's frequency? One well-studied method that people seem to use to judge frequency is influenced by *how easily* a thought or example is brought to mind.

Psychologists describe this manner of judgment of an event's frequency based on how accessible or "available" the event is mentally. If you can easily bring to mind an example of something happening (so the brain's logic goes), it must happen a lot.

This impacts our perception of risk in that if we can easily conjure up a mental image of a negative event happening, we will consider the frequency to be high and the risk to be high. But if we cannot easily conjure up a mental image of a negative event happening, we will consider the frequency to be low and the risk to be low. Indeed, people typically underestimate low-probability risks, such as electrical incidents that account for 0.16% of occupational incidents. And, of course, it is much easier to conjure up mental images of events that you have experienced personally rather than ones that you have merely read about or for which you have only heard statistics about.

Understanding Risk: Feelings as Input

"Affect" is a term used regularly within the field of psychology; a fancy way of referencing the experience of feelings or emotions. Feeling alert, determined, inspired, and proud are examples of "positive" affect, while feeling scared, jittery, ashamed, and irritable are examples of "negative" affect.

As we know, the experiential system (which we usually use to calculate our risk) relies heavily on past experience to influence an individual's perception of risk. A past experience that is laden with affect plays a large role in how the risk is perceived. This is why people who have been emotionally affected by seeing first-hand the effects of a terrible electrical incident do not need much more convincing to be persuaded to be vigilant regarding electrical safety.

When we have a strong emotional experience associated with certain hazards, this has a profound impact on our risk perception. A strong negative experience leads to an increased perception of risk. This concept is certainly not new to science and has been very well captured by the Old English proverb "once bitten, twice shy," originally dating back to English Printer William Caxton in the 15th century.

Essentially, a single experience with strong negative affect can be so easily brought to mind (i.e., becomes so easily *available*) that it is represented to us mentally like a high-frequency event.

To summarize the psychology of risk as relevant to this chapter, we:

1) base our risk calculations primarily on feelings and past experiences;
2) base our perceptions of risk of an event on how easily we can bring an example to mind;
3) have heightened perception of risk for events with which we have an associated strong emotional experience.

These three points provide some of the psychological explanations as to why we "relax" after time goes by without an injury. Because they are integral to how we function psychologically, these points should not be frustrating roadblocks for safety professionals, but rather, keys to consider in developing promising solutions.

Instilling Vulnerability to Counter Cultural Drift

The quote in Box 13.2 is from a plant manager during the course of an investigation of an arc flash incident that resulted in an LTI and long-term disability to a worker. The statement illustrates how the low-frequency nature of electrical injuries can create an inaccurate perception of risk, not just for workers at risk for injury but also for managers who allocate resources to address and mitigate hazards.

As if in response to the questions posed by the manager, the quote in Box 13.3 is from an electrician who had worked for more than 25 years without truly understanding the risk for serious life changing and potentially fatal injury. He described his electrical safety training up to that day as consisting of a review of electrical safety regulations and various employers' safety rules. He admitted he had worked his career without fully understanding the hazards of electricity, although he, his coworkers, and employers had considered him a "qualified" electrician. Through the course of the training on electrical hazards, he had come to realize that his avoidance of electrical injury had been due to luck.

Box 13.3 Example of an Individual Who Has Experienced a Career with Training Focused on Rules and Procedures Which Did Not Address His Personal Vulnerability
I've been in the business (electrical trade) for 25 years. Until today, I can honestly say that I've never understood the hazards I've worked with. – An electrician at the end of an 8-hour hazard awareness seminar

How can we keep people "on guard" when their perception of an electrical incident having low probability is inaccurate? Indeed, this is our intention regarding electrical safety.

"It Could Be Me"

We often try to convince people of their risk by presenting statistics. This is not typically going to be the most persuasive approach because it would require that we calculate our risk via our analytic system. But, as discussed earlier, we are much more likely to use our experiential system. So, a much more persuasive approach would be to present warnings as vivid, emotionally based stories instead of, or at the very least, before presenting statistics. Background stories of people affected by incidents that include pictures with faces, names, and references to personal characteristics (family, hobbies) will persuade much more than simply presenting statistics. This is effective because it makes a person able to *see themselves* represented in the example. The more a person can see him- or herself in the shoes of an example, the more likely he or she is to be affected emotionally by a story. And, the more likely he or she will think, "that could be me." Creating these vignettes is an art; too much detail, and people can no longer see themselves in the story.

Sometimes, however, it is important to present the statistics themselves. In these cases, there are some methods that can help make the statistics more convincing. People are not well-equipped mentally to make good sense of statistics, and the manner in which statistics are presented can have a big impact on how people interpret them. We can utilize this to help heighten (increase) risk perceptions by using the following methods:

1) When possible, prioritize using stories, not statistics. In stories, mention personable information that is generic enough to apply to many people. For example, item B in Box 13.4 mentions the victim's wife and two middle school-aged children – most workers will be able to project themselves into that scenario. However, the information is not so detailed as to be exclusive. That is, the example does not say "two daughters, aged 14 and 17," because that may be too specific to allow others to see themselves in the example.
2) When presenting statistics, present *frequencies* (e.g., 2/10), instead of probabilities (e.g., 20%). A study by Yamagishi found that people rated a disease that killed 1,286 out of 10,000 people to be more dangerous than a disease that killed 24.14% of the population, *even though the actual numbers indicate that the second scenario killed more people.*

3) Present statistics that show risk *over a lifetime*, instead of during a single instance. For example, present to workers the risk of electric shock over a 30-year technical career, not for each individual task, and not for a one-year period. Referring back to Box 13.2, for the example company and based on the private industry electrical fatality rate of 0.25 per 100,000 workers, over the past 20-year period, five electrical fatalities would have occurred, as compared to the actual 0.
4) Regarding manager buy-ins, present statistics in terms of a certain percentage of lives that would be saved *out of* a certain number, instead of just presenting the number of lives saved. For example, Slovic and colleagues found that people judged airplane safety equipment rated to save "98% of 150 lives" to be more worth the investment than equipment rated to save "150 lives." Why? It was hypothesized that proving only the total number weakens our ability to evaluate the benefit, whereas providing the *percentage saved out of that number* strengthens our ability to evaluate the benefit.

Box 13.4 Two Hypothetical Examples That Could Be Used in a Company Newsletter. Which of These Two Examples Is More Likely to Persuade Your Workers to Take Appropriate Safety Measures?

A) "There are 1,890 electrical shock and burn injuries each year."
B) "Last Friday, one of our linemen was critically injured in an electric shock incident. His wife and their two middle school children have spent the week visiting him in the hospital and would like to request that get-well cards be sent to the following address ..."

What You Can Do

> The only constant is change, continuing change, inevitable change, which is the dominant factor in society today. No sensible decision can be made any longer without taking into account not only the world as it is, but the world as it will be.
>
> –Isaac Asimov

Asimov's statement implies that no matter where an organization's electrical safety culture may be centered, it is not static. The danger is that it could be drifting in a negative way, as described in the definition of cultural drift at the beginning of this chapter. Sustaining a strong electrical safety culture requires deliberate continuous effort. Here are some examples that may be helpful:

1) Educate yourself, coworkers, and managers on the low-frequency/high-consequence characteristic of electrical injuries.

    Seek first to understand … and then to be understood.
    –Steven Covey

    In *7 Habits of Highly Successful People*, Covey used this statement to emphasize exploring other people's needs and concerns before imposing your views on them. Here, we propose a slightly different application. Use the data and models in this chapter to build the understanding that electrical injuries are relatively rare and that the consequences of injury are among the most severe – first for your own understanding and then to help educate others. Study the work of Peterson, Manuele, Krause, and others who pioneered and continue to advance research and practice in safety management techniques targeting low-frequency/high-consequence injuries. Use what you learn to influence how others think about their vulnerability to electrical hazards.
2) Eliminate characterizations that trivialize electrical injuries, i.e., using terms for electric shocks as "minor shocks," "tingles," or "nips." These terms instill a perception that the events do not have high-potential consequences. "Nonfatal" shock would be more accurate and is a reminder of the fatality potential.
3) Conceptualize incidents that do occur, not as sporadic or random, but as indicators of a safety program or culture that needs to be improved.
4) Investigate incidents that occurred that have high potential for fatal injury, but did not result in injury. Use these examples as learning tools for workers at risk, supervisors, safety professionals, managers, design engineers, etc. These incidents, whether called near-miss or near-hits, provide invaluable insight into underlying deficiencies in electrical safety management. Recognizing the high-consequence risk potential can help sustain a sense of vulnerability critical to maintaining a strong culture of prevention.
5) Promote awareness of consequences, other than injury, of mishaps involving interactions between people and electrical hazards. Disruption of electrical energy, control, communications, and data systems is a more likely consequence compared to actual injury, but understanding the underlying causes can help maintain support for electrical safety programs. The work to prevent incidents and mishaps in electrical systems critical to operations prevents not only injuries but also disruption and damage with broad and costly implications. Understanding this may help maintain support for electrical safety even in the absence of injuries.

6) Build a strong culture of first- and second-party audits that assess compliance with practices, procedures, and processes designed to prevent electrical incidents. These audits should include assessment of understanding of risk and not just compliance with rules. For example, in addition to assessing compliance to use GFCI protection on certain work activities ("following the rules"), also assess workers' and supervisors' understanding that an incident around electricity has a high potential for severe injury or fatality and how the GFCI mitigates that risk (understanding vulnerability – "it could be me").
7) Establishing an awareness of vulnerability in employees should be a prerequisite to training in rules, regulations, and procedures. Studies have shown that a combination of induced vulnerability or threat *with* the self-efficacy (i.e., competency) to avoid the hazard is one of the ideal methods in persuading people to engage in self-protective behavior. In other words, a threatening message can be provided and should then be followed by instructions on ways to avoid that threat.
8) Establish an electrical safety management process based on continuous improvement, using the Plan-Do-Check-Act quality improvement model attributed to D. Edwards Deming. It is embedded in essentially every business management improvement process. It is simple and powerful when executed with discipline and rigor. It is an essential component of *ANSI Z10 Occupational Health and Safety Management Systems* and ISO 455001 *Occupational Health and Safety Management Systems – with Requirements for Use.*

There Is No End

Electrical injury and fatality frequency are relatively small numbers within the context of total injury and fatality frequency. However, the severity of electrical injury in terms of emotional and physical damage, as well as medical cost to employers, the damage to critical infrastructure, and disruption to operations are significantly higher than the most common causes of injuries in the workplace (e.g., musculoskeletal disorders).

The fact that electrical incidents are low in frequency can do two things:

1) Create the false perception that the management of the hazard is under control.
2) Create a challenge for workers to stay vigilant in the face of a low probability of occurrence.

These two results are not appropriate for a hazard with such severe consequence potential. Because we mentally develop our risk perceptions based on perceived frequencies, the low frequency of electrical injuries lowers the perception of the risk of electrical injury such that we lower our guard and do not stay vigilant.

However, because we also perceive risk based on our emotional perception (affect), safety professionals can use this to our advantage in how we teach people about risk.

In order to stay vigilant against incidents, we cannot only talk about electrical safety as rules, regulations, and standards. To prevent cultural drift, we have to raise awareness of our vulnerability to fatality or debilitating injury and to the related significant medical cost. To do that, we need to examine how we talk and act that helps create the perception of this risk.

The beginning of this chapter posed some questions about how electrical safety programs are impacted when people involved with them have never seen a serious injury or fatality firsthand. Could the absence of serious injuries contribute to a *cultural drift* with a subsequent *occlusion* of electrical safety? How can we maintain an *organizational culture* around electrical safety built on motivation, passion, and understanding of vulnerability without witness to serious injury or fatality? The answer here is complex. Part of the answer is two-fold: that our workers and management must be able to (1) maintain high-risk perceptions despite the low frequency of electrical injuries and (2) believe that "It could happen to me"; that is, they should be able to see themselves (or their employees) in the shoes of the stories and statistics of injury. If leaders in electrical safety can perfect methods and their application to achieve these solutions, it is possible to avoid the electrical fatalities and serious injuries that could possibly result from cultural drift.

The end game is fact, continual improvement in identifying hazards, applying risk controls, evaluating the effectiveness of applied controls, and assessing residual risk for opportunities for additional controls.

There is no end.

Crosstalk: Questions for Discussion

1 What is the process to review, update, and communicate changes to your electrical safety program?

2 Who should be involved in an annual management review of electrical safety performance?

3 What motivates workers, supervisors, and managers to be engaged in electrical safety? How do you feel about your answer?

4 What does continual improvement in electrical safety mean to you?

5 How do you define a near-miss? How does your organization describe a near-miss? What is the process used to learn from near-misses?

References

1 Report of the Columbia Accident Review Board (2003). 6 volumes, vol. 1, p. 97. Government Printing Office, Washington, D.C. http://caib.nasa.gov/.

2 "What Is Cultural Drift? How To Keep Workplace Progress From Slipping" By Dr. Diane Hamilton, www.forbes.com, May 25, 2025

Further Reading

Anderson, M. and Denkl, M. The Heinrich triangle – too simplistic a model for HSE management in the 21st century? *Society of Petroleum Engineers 2010 International Conference on Health Safety and Environment.*

ANSI/ASSP Z10. Occupational health and safety management systems, American Society of Safety Professionals.

Casini, V., Stout, N.A and Kisner, S. Worker deaths by electrocution, *National Institute for Occupational Safety and Health, publication no. 98–131* (May 1998).

Cawley, J.C. and Brenner, B.C. Occupational electrical injuries in the U.S., 2003–2009. *2012 IEEE IAS Electrical Safety Workshop* (30 January–3 February 2012), Daytona Beach, FL.

Cawley, J.C. and Brenner, B.C. 20 years of electrical injury data shows substantial improvement. www.esfi.org (accessed 14 December 2012).

Cawley, J.C., and Homce, G. Trends in U.S. electrical injuries 1992–2002. *IEEE Transactions on Industry Applications*, Aug/Sept 2008.

Caxton, W. 1484, retrieved from Wiki Quote http://en.wikiquote.org/wiki/English_proverbs#O.

Decker, S., "Safety Theater: How success can mask growing safety risks", *Professional Safety*, vol. 70, no. 10, Oct 2025, pp 28–34.

Doughty, R.L., Epperly, R.A. and Jones, R.A., "Maintaining safe electrical work practices in a competitive environment", *IEEE Transactions on Industry Applications*, vol. 28, no. 1, 1992, pp 196–204.

Finucane, M.L., Alhakami, A., Slovic, P., and Johnson, S.M. (2000). The affect heuristic in judgments of risks and benefits. *Journal of Behavioral Decision Making* 13 (1): 1–17.

Floyd, H.L. Lessons in assuring a sustainable program. 2012 *International Maintenance Conference* (3–7 December 2012), Bonita Springs, Florida

Floyd, H.L., Cole, B.C. Twenty years later: creating a continuous improvement environment for electrical safety, scheduled for presentation at *2013 IEEE IAS Petroleum and Chemical Industry Conference* (23–25 September 2013), Chicago, IL

Floyd, H.L., Cole, B.C., Doughty, R.L. et al. (1994). Creating a continuous improvement environment for electrical safety. *IEEE Transactions on Industry Applications* 30 (3): 543–552.

Gigerenzer, G., Gaissmaier, W., Kurz-Milcke, E. et al. (2008). Helping doctors and patients make sense of health statistics. *Psychological Science in the Public Interest* 8 (2): 53–96.

Heinrich, H.W., Peterson, D., and Roos, N. (1980). *Industrial Accident Prevention – A Safety Management Approach*. New York: McGraw Hill.

Hendrickx, L., Vlek, C., and Oppewal, H. (1989). Relative importance of scenario information and frequency information in the judgment of risk. *Acta Psychologica* 72: 41–63.

Keller, C., Siegrist, M., and Gutscher, H. (2006). The role of the affect and availability heuristics in risk communication. *Risk Analysis* 26 (3): 631–639.

Krause, T.R. A new paradigm for fatality prevention. *Safety* 2010 *Conference, American Society of Safety Engineers* (13–15 June 2010), Baltimore, MA.

Levenson, N., Dulac, N., Barrett, B., et al. Risk analysis of NASA independent technical authority. *Mass. of Technology*, June 2005, p. 5.

Lin, M.-L. and Solomon, P. (2012). Tales of the unexpected. *Safety and Health Practitioner* 37–40. Figure 8 © DuPont, used with permission.

Manuele, F. (2008). *Advanced Safety Management*, 84. Wiley.

Manuele, F.A. (2008). Serious injuries & fatalities – a call for a new focus on their prevention. *Journal of Professional Safety*, American Society of Safety Engineers 53: 32–39.

Slovic, P., Fischhoff, B., and Lichtenstein, S. (1978). Accident probabilities and seat belt usage: a psychological perspective. *Accident; Analysis and Prevention* 10: 281–285.

Slovic, P., Finucane, M.L., Peters, E., and MacGregor, D.G. (2002). The affect heuristic. In: *Heuristics and Biases: The Psychology of Intuitive Judgment* (ed. T. Gilovich, D. Griffin, and D. Kahneman), 397–420. New York: Cambridge University Press.

Slovic, P., Finucane, M.L., Peters, E., and MacGregor, D.G. (2004). Risk as analysis and risk as feelings: some thoughts about affect, reason, risk, and rationality. *Risk Analysis* 24 (2): 311–322.

Tversky, A. and Kahneman, D. (1982). Availability: a heuristic for judging frequency and probability. In: *Judgment under Uncertainty: Heuristics and Biases* (ed. D. Kahneman, P. Slovic, and A. Tversky), 163–178. Cambridge UK: Cambridge University Press.

Van'T Reit, J., Ruiter, R.A.C., Werrij, M.Q., and De Vries, H. (2008). The influence of self-efficacy on the effects of framed health messages. *European Journal of Social Psychology* 38: 800–809.

Watson, D. and Clark, L.A. (1994). THE PANAS-X: manual for the positive and negative affect schedule – expanded form, retrieved online August 10, 2012 from www.psychology.uiowa.edu/faculty/clark.

Weinstein, N.D. (1989). Effects of personal experience on self-protective behavior. *Psychological Bulletin* 105: 31–50.

Wyzga, R.E. and Lindroos, W. (1999). Health implications of global electrification. *Annals of the New York Academy of Sciences* 888: 1–7.

Yamagishi, K. (1997). When a 12.86% mortality is more dangerous than 24.14%: implications for risk communication. *Applied Cognitive Psychology* 11: 495–506.

Appendix: Occupational Electrical Injury and Fatality Statistics

Credible data is essential in telling the story of electrical safety. There may be more than those discussed here, but three sources of credible detailed data for electrical injuries and fatalities are the Fire Protection Research Foundation, CPWR – The Center for Construction Research and Training, and Electrical Safety Foundation International. These three organizations periodically analyze data from the US Bureau of Labor Statistics (BLS) and other credible sources and publish reports that are useful in identifying opportunities to improve electrical safety efforts. Their reports can be downloaded free of charge.

National Fire Protection Foundation

https://www.nfpa.org/Education-and-Research/Research/Fire-Protection-Research-Foundation.

The research affiliate of the National Fire Protection Association and the National Fire Prevention Foundation addresses many different aspects of electrical safety. There are nearly 20 different reports available. One that specifically discusses occupational injuries is "Occupational Injuries from Electrical Shock and Arc Flash Events" by Richard B. Campbell, ScD, NFPA, and David A. Dini, P.E., UL, LLC published March 1, 2015.

The Center for Construction Research and Training

https://www.cpwr.com/.

CPWR is a nonprofit created by North America's Building Trades Unions (NABTU) and dedicated to reducing occupational injuries, illnesses, and fatalities in the construction industry. CPWR has been a partner with the NIOSH since NIOSH launched its construction safety and health research initiative in 1990. Free resources are available at https://www.cpwr.com/.

A recent analysis of electrical injuries and fatalities in construction is *Electrical Injuries and Citations in Construction,* CPWR Data Bulletin, November 2022.

Electrical Safety Foundation International (ESFI)

https://www.esfi.org/.

Hidden Risk in Occupational Electrical Safety: Through the Lens of System Safety,
First Edition. H. Landis "Lanny" Floyd.

ESFI is the premier nonprofit organization dedicated exclusively to promoting electrical safety at home and in the workplace. Founded in 1994 as a cooperative effort by the National Electrical Manufacturers Association (NEMA), Underwriters Laboratories (UL), and the US Consumer Product Safety Commission (CPSC). ESFI is dedicated exclusively to promoting electrical safety at home and in the workplace through education, awareness, and advocacy.

- ESFI partners with corporations, industry organizations, community leaders, and the media to develop key resources and consistent electrical safety messages.
- ESFI proudly sponsors National Electrical Safety Month each May and engages in public education campaigns throughout the year to prevent electrical fires, injuries, and fatalities. We encourage and support organizations in their efforts to promote electrical safety.
- ESFI educates consumers and the public about how to recognize warning signs of electrical hazards, providing information about new, advanced safety technology that offers enhanced protection in the home.
- ESFI promotes safety in the workplace by providing dynamic educational and awareness materials to help employers ensure a safe working environment for their employees.
- ESFI provides vital third-party validation by advocating the adoption of legislation, codes, and standards that are consistent with our mission of electrical safety.
- ESFI serves as a national resource and the leading authority for electrical safety information. We are often called upon by media outlets to serve as experts on electrical safety issues.

2011–2022 Data Summary

The Electrical Safety Foundation has compiled occupational electrical injury and fatality data from the BLS for the period 2011–2023. A total of 70,692 occupational fatalities occurred from all causes. Of these, 1,940 were due to contact with electricity.

98.9% of the electrical fatalities in this period were men. The percentage of fatalities that involved self-employed workers was similar in both the overall and electrically related workplace fatality data (20% and 21%, respectively). 64% of overall workplace fatalities involved White workers, 19% involved Hispanic or Latino workers, and 11% involved Black workers. For electrically related fatalities, these figures were 63%, 28%, and 6%, respectively.

The construction industry had the highest number of electrical fatalities (855), followed by professional and business services (212), trade, transportation, and utilities (155), natural resources and mining (138), and manufacturing (120).

Just five occupations in the construction trades – electricians, construction laborers, roofers, painters, and carpenters – experienced 30% of all electrical fatalities.

Among the workers who were fatally injured by electricity, 33% died at a private residence, 28% on industrial premises, and 11% on a street or highway. 96% were employed in private industry.

ESFI compiles data from the BLS and the OSHA to track electrical safety trends in the workplace. Its report, published in 2025, which analyzed data from 2011 to 2022, includes the information below. More information and resources are available on its website.

- Contact with electricity is one of the leading causes of fatalities in the workplace.
- 74% of workplace electrical fatalities occurred in nonelectrical occupations according to OSHA.
- 150 workplace electrical fatalities occur on average every year according to the BLS.
- 5.3% of all workplace fatalities were caused by contact with electricity.
- Electrical fatality rates have remained consistent year over year. More needs to be done to reduce the amount of electrical fatalities in the workplace.
- Hispanic or Latino workers have the highest rate of electrical fatalities and experience a disproportionate amount of electrical fatalities.
- Construction and extraction occupations; installation, maintenance, and repair occupations; and building and grounds cleaning and maintenance occupations have the highest rate of electrical fatalities.

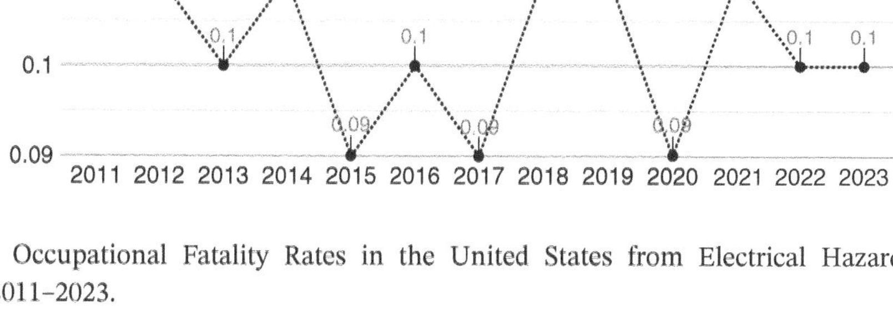

Occupational Fatality Rates in the United States from Electrical Hazards 2011–2023.

US occupational fatality rate per 100,000 workers by select occupations 2011–2023.

Occupation	Rate
Electrical power line installers and repairers	6.56
Electricians	2.75
Roofers	2.66
Grounds maintenance workers	1.24
Heating, air conditioning, and refrigeration mechanics	0.97
Construction laborers	0.55
Painters, construction, and maintenance	0.41
All occupations	0.10

Ten occupations accounted for over 58% of all workplace electrical fatalities:

1) Electricians
2) Laborers, except construction
3) Construction laborers
4) Electrical power installers and repairers
5) Tree trimming occupations
6) Electricians' apprentices
7) Heating, air conditioning, and refrigeration mechanics
8) Roofers
9) Heavy truck drivers
10) Painters in construction or maintenance

Occupations Having the Most Electrical Fatalities in the United States, 2011–2023

91% of all workplace electrical fatalities were caused by:

1) Overhead power line contact: 42.8%
2) Unexpected contact with electricity: 19.3%
3) Nearby energized equipment contact: 12.7%
4) Working on energized parts: 4.1%
5) Ground faults: 4.0%
6) Damaged wiring or equipment: 3.1%

Top Causes of Occupational Electrical Fatalities in the United States, 2011–2023

US occupational electrical fatality rate per 100,000 workers by select industry, 2011–2023.

Industry	Rate
Construction	0.68
Natural resources and mining	0.32
Professional and business services	0.11
All occupations	0.10
Manufacturing	0.06
Trade, transportation, and utility	0.04

2003–2010 Data Summary

ESFI's occupational electrical injury and fatality information has been compiled from data published by the BLS and the OSHA.

Each year, the BLS performs its Census of Fatal Occupational Injuries (CFOI). This is an actual count, or census, of fatal injuries. Each case is verified by two or more independent sources of information. Such sources can include death certificates, police reports, news reports, OSHA reports, etc. Similarly, to estimate the number of nonfatal injuries and illnesses, the BLS performs its Survey of Occupational Injuries and Illnesses. Nonfatal injuries, due to their sheer number, are statistically estimated based on a large annual survey of injuries reported by employers.

OSHA 170 reports provide additional context regarding the incidents, including occupation, cause, and a narrative of what led to the electrical incident. OSHA 170 forms are made available to the public once the report has been completed and personal information is removed from the report. This may result in slightly different totals between the two data sources.

The Electrical Safety Foundation has compiled the occupational electrical injury experience of the major industries and occupations from data available through the BLS for the period 2003–2010. A total of 42,882 occupational fatalities occur from all causes, of which 1,738 were due to contact with electric current.

The construction industry had the highest number of electrical fatalities (849), followed by professional and business services (208), trade transportation and utilities (182), natural resources good morning (154), and manufacturing (137). Just five occupations in the construction trades – electricians, construction laborers, roofers, painters, and carpenters – experience more than 32% of all electrical fatalities, electrical power line installers and repairs about 8%, and tree trimmers about 5%. All of the 163 electrical fatalities during 2010 were men. The self-employed were about 22% of all occupational deaths, but only 19% of electrical deaths. Nearly 68% were White, less than 6% were Black, and 24% were Hispanic. 98% died of electrocution, 63% were constructing, repairing, or cleaning something at the time of death, 34% died on industrial premises, 28% at a private residence and nearly 12% on a street or highway, and 96% were employed in private industry. In order to fairly compare industries and occupations with different numbers of employees in parentheses, different total exposures to electrical hazards in parentheses rates of fatal and nonfatal electrical injury were computed. It was shown that electrical fatalities for approximately 4% of all occupational fatalities each year between 2003 and 2010 contact with overhead power lines was the leading fatal injury event for the period, but was a minor source of nonfatal electrical injury. Contact with wiring, transformers, or other electrical components; contact with the electric current of machines, tools, appliances, or light fixtures; and contact with electric current unspecified where the next largest fatal event categories, respectively. Workers in four industries – utilities, mining, construction, agricultural, forestry, fishing, and hunting experienced electrical fatality rates in excess of the private industry rate every year between 2003 and 2010. Utility and construction workers also experience nonfatal electrical injury rates in excess of the private industry rate each year between 2003 and 2010. Read more in our resource *Electrical Safety Then and Now – Twenty Years of Electrical Injury Data*, which shows substantial electrical safety improvement and discusses some of the more interesting patterns observed in occupational electrical injury and fatality statistics for 1992–2010.

Electrical Safety Then and Now 1992–2010

Twenty Years of Electrical Injury Data Shows Substantial Electrical Safety Improvement

The BLS first began to compile its CFOI in 1992. This actual count on-the-job injuries and illnesses has since been accepted by the NSC as the definitive census of occupational injuries in the United States. Each fatality is verified by at least two authoritative sources, such as death certificates, police reports, and newspaper articles. In addition, the BLS also estimates the annual number of nonfatal injuries to US workers through the Survey of Occupational Injury and Illnesses (SOII). The SOII is a statistical estimate of nonfatal injuries based on a survey of more than 200,000 employers each year.

ESFI uses CFOI and SOII to distill information specifically pertaining to fatal and nonfatal occupational electrical injuries. Each year the ESFI publishes electrical injury information in tabular and graphical form on its website. This article discusses some of the more interesting patterns observed over the 20-year period from 1992 to 2010.

Since 1992, fatal electrical injuries have declined by more than 50% on an annual basis and that trend has accelerated since 2006. For the 15 years between 1992 and 2006, there was a 25% decline in annual electrical fatalities, but in just five years from 2006 to 2010, electrical fatalities declined by one third (Figure A.1).

"Contact with Overhead Power Lines" (includes direct worker contact and contact with machines, tools, and hand-carried metallic objects) was the largest fatal accident category, comprising 40–44% of all electrical fatalities for 1992–2010. "Contact with Wiring, Transformers, or Other Electrical Components" (most common for workers who install repair and maintain electrical systems and apparatus in the normal course of their electrical work) was the second largest fatal injury category with 20% of all fatalities, followed by "Contact with the Electric Current of Machine, Appliance, or Light Fixture" (most common for workers in the normal course of their nonelectrical work) with 17%. Even though the

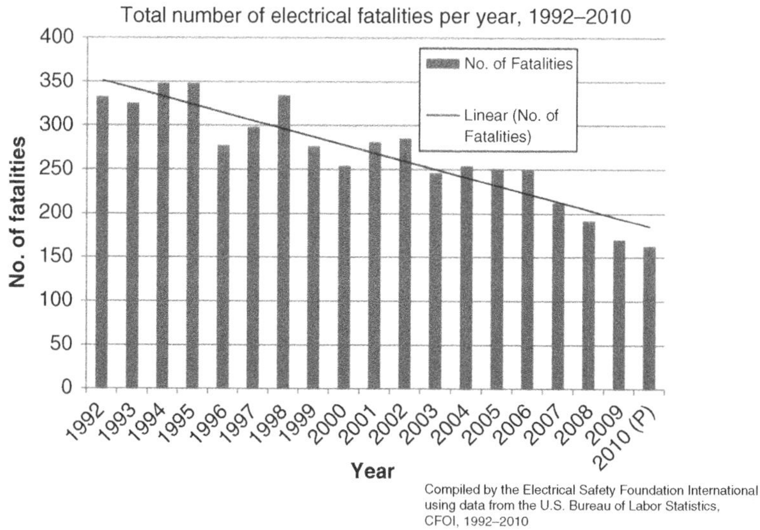

Figure A.1 Annual electric fatalities declined by more than 50% for the period 1992–2010. Adapted from U.S. Bureau of Labor Statistics, CFOI, 1992–2010.

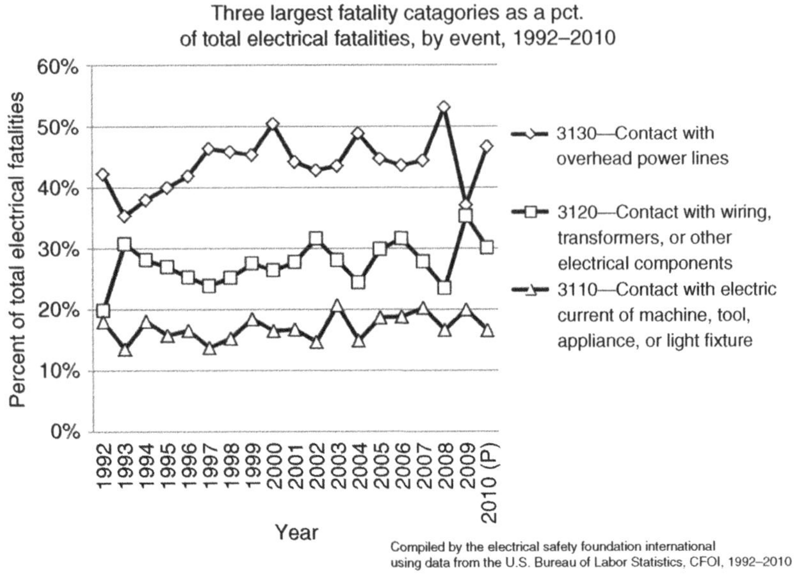

Figure A.2 The percentage contribution of the three largest categories of fatal electrical accidents remained relatively unchanged despite a 50% decline in the total number of fatal electrical accidents. pct.: percentage. Adapted from U.S. Bureau of Labor Statistics, CFOI, 1992-2010.

number of fatal accidents declined by 50%, the relative percentage contribution of the three largest accident categories remained stable (Figure A.2).

Fatal electrical injuries have shown an even more dramatic improvement. In 2010, nonfatal electrical injuries were down by more than 60% from their 1992 level. The largest categories of nonfatal electrical injury are "Contact with the electric current of a machine, tool, appliance, or light fixture" and "Contact with wiring, transformers, or other electrical components." Figure A.3 shows that the relative contribution of the two largest categories has remained relatively unchanged despite the 60% decline in total nonfatal electrical accidents. Just five occupational groups account for nearly 80% of all fatal electrical accidents. Construction trade workers alone represent about 38% of all electrical fatalities. Other occupational groups with high numbers of electrical fatalities are "Installation, Maintenance, and Repair Occupations" (21%), "Grounds Maintenance Workers (7%), Transportation and Moving Materials Occupations" (6%), "Other Management Occupations (4%), and Agricultural Workers" (2%).

The construction industry experienced 52% of total electrical fatalities and far outpaced all others (Figure A.4).

BLS data shows that as few as one to four electrical fatalities annually were attributed to electrical burns for 2003–2010. Some electrical experts consider this

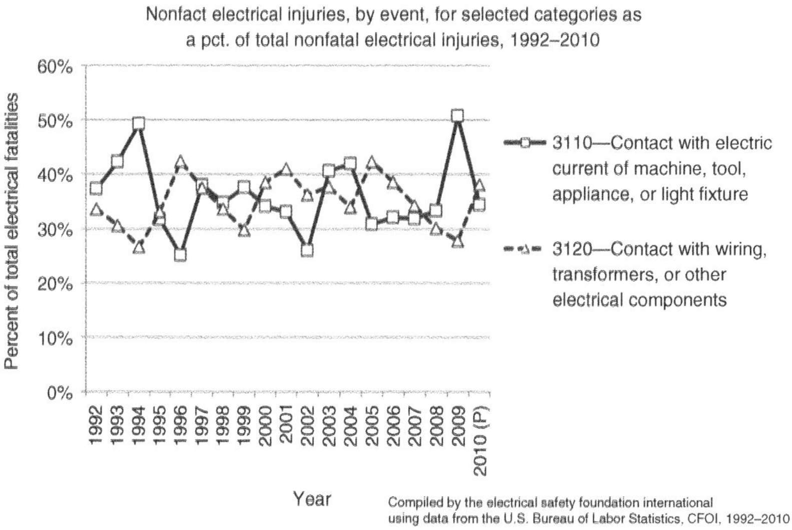

Figure A.3 Nonfatal electrical injuries as a percentage of the total for the two largest categories of electrical accidents remained relatively unchanged for 1992–2010, even though the total number of nonfatal electrical accidents declined by more than 60%. Adapted from U.S. Bureau of Labor Statistics, CFOI, 1992–2010.

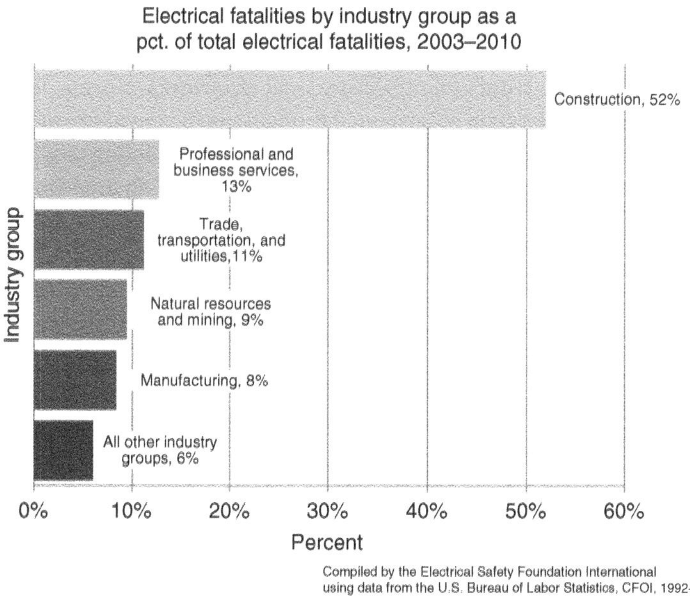

Figure A.4 The construction industry alone sustained 52% of all fatal electrical accidents for the period 2003–2010. Adapted from U.S. Bureau of Labor Statistics, CFOI, 1992–2010.

to be an undercount. About 39% of nonfatal electrical injuries are electrical burns. In the construction industry, about 57% of electrical injuries are burns. The utility industry is the only other in which nonfatal burn injuries outnumber electric shock injuries (76%). The utility industry has the highest rate of nonfatal electrical burn injury at 1.6 cases per 10,000 workers in 2010, followed by the construction industry at 0.4 cases. The overall electrical burn injury rate for private industry remained at 0.1 cases per 10,000 workers for 2003–2010.

The construction industry, however, had the highest rate of nonfatal electric shock injuries at 0.6 cases per 10,000 workers, whereas the utility industry improved from 0.7 cases in 2009 to 0.4 cases in 2010; the overall electric shock rate for private industry remained at 0.2 cases per 10,000 workers for 2003–2010.

Since 1992, both fatal and nonfatal electrical injuries have shown significant and sustained declines. The recent slowdown in economic activity has probably contributed to the even sharper declines in electrical accidents over the last few years. "Contact with Overhead Power Lines" remains a significant problem, accounting for nearly one-half of all occupational electrical fatalities. The fatal data from 1992 to 2010 shows that although the number of electrical fatalities has decreased, the overall mix of fatal accidents remains largely unchanged. The same can be said for nonfatal electrical accidents. The construction and utility industries remain problem areas in terms of both fatal and nonfatal electrical accidents. Although the construction industry sustains a much larger number of electrical injuries than the utility industry, the utility industry exhibits a higher rate of both fatal and nonfatal electrical injuries.

The advancements of NFPA 70E is an important electrical safety standard that is surely a component of the reduction of occupational electrical accidents. Real improvement in electrical safety can be sustained through the increased use of the technique because the methods found in NFPA 70E and through training targeted people in high-risk occupations and industries electrical safety is an area where perseverance pays off.

Index

Hidden Risk in Occupational Electrical Safety: Through the Lens of System Safety,
First Edition. H. Landis "Lanny" Floyd.

IEEE Press Series on Power and Energy Systems

Series Editor: Ganesh Kumar Venayagamoorthy, Clemson University, Clemson, South Carolina, USA.

The mission of the IEEE Press Series on Power and Energy Systems is to publish leading-edge books that cover a broad spectrum of current and forward-looking technologies in the fast-moving area of power and energy systems including smart grid, renewable energy systems, electric vehicles and related areas. Our target audience includes power and energy systems professionals from academia, industry and government who are interested in enhancing their knowledge and perspectives in their areas of interest.

1. *Electric Power Systems: Design and Analysis, Revised Printing*
   Mohamed E. El-Hawary
2. *Power System Stability*
   Edward W. Kimbark
3. *Analysis of Faulted Power Systems*
   Paul M. Anderson
4. *Inspection of Large Synchronous Machines: Checklists, Failure Identification, and Troubleshooting*
   Isidor Kerszenbaum
5. *Electric Power Applications of Fuzzy Systems*
   Mohamed E. El-Hawary
6. *Power System Protection*
   Paul M. Anderson
7. *Subsynchronous Resonance in Power Systems*
   Paul M. Anderson, B.L. Agrawal, and J.E. Van Ness
8. *Understanding Power Quality Problems: Voltage Sags and Interruptions*
   Math H. Bollen
9. *Analysis of Electric Machinery*
   Paul C. Krause, Oleg Wasynczuk, and S.D. Sudhoff
10. *Power System Control and Stability, Revised Printing*
    Paul M. Anderson and A.A. Fouad

11. *Principles of Electric Machines with Power Electronic Applications, Second Edition*
Mohamed E. El-Hawary
12. *Pulse Width Modulation for Power Converters: Principles and Practice*
D. Grahame Holmes and Thomas Lipo
13. *Analysis of Electric Machinery and Drive Systems, Second Edition*
Paul C. Krause, Oleg Wasynczuk, and S.D. Sudhoff
14. *Risk Assessment for Power Systems: Models, Methods, and Applications*
Wenyuan Li
15. *Optimization Principles: Practical Applications to the Operations of Markets of the Electric Power Industry*
Narayan S. Rau
16. *Electric Economics: Regulation and Deregulation*
Geoffrey Rothwell and Tomas Gomez
17. *Electric Power Systems: Analysis and Control*
Fabio Saccomanno
18. *Electrical Insulation for Rotating Machines: Design, Evaluation, Aging, Testing, and Repair*
Greg C. Stone, Edward A. Boulter, Ian Culbert, and Hussein Dhirani
19. *Signal Processing of Power Quality Disturbances*
Math H.J. Bollen and Irene Y.H. Gu
20. *Instantaneous Power Theory and Applications to Power Conditioning*
Hirofumi Akagi, Edson H. Watanabe, and Mauricio Aredes
21. *Maintaining Mission Critical Systems in a 24/7 Environment*
Peter M. Curtis
22. *Elements of Tidal-Electric Engineering*
Robert H. Clark
23. *Handbook of Large Turbo-Generator Operation and Maintenance, Second Edition*
Geoff Klempner and Isidor Kerszenbaum
24. *Introduction to Electrical Power Systems*
Mohamed E. El-Hawary
25. *Modeling and Control of Fuel Cells: Distributed Generation Applications*
M. Hashem Nehrir and Caisheng Wang

26. *Power Distribution System Reliability: Practical Methods and Applications*
Ali A. Chowdhury and Don O. Koval

27. *Introduction to FACTS Controllers: Theory, Modeling, and Applications*
Kalyan K. Sen and Mey Ling Sen

28. *Economic Market Design and Planning for Electric Power Systems*
James Momoh and Lamine Mili

29. *Operation and Control of Electric Energy Processing Systems*
James Momoh and Lamine Mili

30. *Restructured Electric Power Systems: Analysis of Electricity Markets with Equilibrium Models*
Xiao-Ping Zhang

31. *An Introduction to Wavelet Modulated Inverters*
S.A. Saleh and M.A. Rahman

32. *Control of Electric Machine Drive Systems*
Seung-Ki Sul

33. *Probabilistic Transmission System Planning*
Wenyuan Li

34. *Electricity Power Generation: The Changing Dimensions*
Digambar M. Tagare

35. *Electric Distribution Systems*
Abdelhay A. Sallam and Om P. Malik

36. *Practical Lighting Design with LEDs*
Ron Lenk and Carol Lenk

37. *High Voltage and Electrical Insulation Engineering*
Ravindra Arora and Wolfgang Mosch

38. *Maintaining Mission Critical Systems in a 24/7 Environment, Second Edition*
Peter Curtis

39. *Power Conversion and Control of Wind Energy Systems*
Bin Wu, Yongqiang Lang, Navid Zargari, and Samir Kouro

40. *Integration of Distributed Generation in the Power System*
Math H. Bollen and Fainan Hassan

41. *Doubly Fed Induction Machine: Modeling and Control for Wind Energy Generation Applications*
Gonzalo Abad, Jesús López, Miguel Rodrigues, Luis Marroyo, and Grzegorz Iwanski

42. *High Voltage Protection for Telecommunications*
Steven W. Blume

43. *Smart Grid: Fundamentals of Design and Analysis*
James Momoh

44. *Electromechanical Motion Devices, Second Edition*
Paul Krause Oleg, Wasynczuk, and Steven Pekarek

45. *Electrical Energy Conversion and Transport: An Interactive Computer-Based Approach, Second Edition*
George G. Karady and Keith E. Holbert

46. *ARC Flash Hazard and Analysis and Mitigation*
J.C. Das

47. *Handbook of Electrical Power System Dynamics: Modeling, Stability, and Control*
Mircea Eremia and Mohammad Shahidehpour

48. *Analysis of Electric Machinery and Drive Systems, Third Edition*
Paul C. Krause, Oleg Wasynczuk, S.D. Sudhoff, and Steven D. Pekarek

49. *Extruded Cables for High-Voltage Direct-Current Transmission: Advances in Research and Development*
Giovanni Mazzanti and Massimo Marzinotto

50. *Power Magnetic Devices: A Multi-Objective Design Approach*
S.D. Sudhoff

51. *Risk Assessment of Power Systems: Models, Methods, and Applications, Second Edition*
Wenyuan Li

52. *Practical Power System Operation*
Ebrahim Vaahedi

53. *The Selection Process of Biomass Materials for the Production of Bio-Fuels and Co-Firing*
Najib Altawell

54. *Electrical Insulation for Rotating Machines: Design, Evaluation, Aging, Testing, and Repair, Second Edition*
Greg C. Stone, Ian Culbert, Edward A. Boulter, and Hussein Dhirani

55. *Principles of Electrical Safety*
Peter E. Sutherland

56. *Advanced Power Electronics Converters: PWM Converters Processing AC Voltages*
Euzeli Cipriano dos Santos Jr. and Edison Roberto Cabral da Silva

57. *Optimization of Power System Operation, Second Edition*
Jizhong Zhu

58. *Power System Harmonics and Passive Filter Designs*
J.C. Das

59. *Digital Control of High-Frequency Switched-Mode Power Converters*
Luca Corradini, Dragan Maksimoviæ, Paolo Mattavelli, and Regan Zane

60. *Industrial Power Distribution, Second Edition*
Ralph E. Fehr, III

61. *HVDC Grids: For Offshore and Supergrid of the Future*
Dirk Van Hertem, Oriol Gomis-Bellmunt, and Jun Liang

62. *Advanced Solutions in Power Systems: HVDC, FACTS, and Artificial Intelligence*
Mircea Eremia, Chen-Ching Liu, and Abdel-Aty Edris

63. *Operation and Maintenance of Large Turbo-Generators*
Geoff Klempner and Isidor Kerszenbaum

64. *Electrical Energy Conversion and Transport: An Interactive Computer-Based Approach*
George G. Karady and Keith E. Holbert

65. *Modeling and High-Performance Control of Electric Machines*
John Chiasson

66. *Rating of Electric Power Cables in Unfavorable Thermal Environment*
George J. Anders

67. *Electric Power System Basics for the Nonelectrical Professional*
Steven W. Blume

68. *Modern Heuristic Optimization Techniques: Theory and Applications to Power Systems*
Kwang Y. Lee and Mohamed A. El-Sharkawi

69. *Real-Time Stability Assessment in Modern Power System Control Centers*
Savu C. Savulescu

70. *Optimization of Power System Operation*
Jizhong Zhu

71. *Insulators for Icing and Polluted Environments*
Masoud Farzaneh and William A. Chisholm

72. *PID and Predictive Control of Electric Devices and Power Converters Using MATLAB®/Simulink®*
Liuping Wang, Shan Chai, Dae Yoo, Lu Gan, and Ki Ng

73. *Power Grid Operation in a Market Environment: Economic Efficiency and Risk Mitigation*
Hong Chen

74. *Electric Power System Basics for Nonelectrical Professional, Second Edition*
Steven W. Blume

75. *Energy Production Systems Engineering Thomas*
Howard Blair

76. *Model Predictive Control of Wind Energy Conversion Systems*
Venkata Yaramasu and Bin Wu

77. *Understanding Symmetrical Components for Power System Modeling*
J.C. Das

78. *High-Power Converters and AC Drives, Second Edition*
Bin Wu and Mehdi Narimani

79. *Current Signature Analysis for Condition Monitoring of Cage Induction Motors: Industrial Application and Case Histories*
William T. Thomson and Ian Culbert

80. *Introduction to Electric Power and Drive Systems*
Paul Krause, Oleg Wasynczuk, Timothy O'Connell, and Maher Hasan

81. *Instantaneous Power Theory and Applications to Power Conditioning, Second Edition*
Hirofumi, Edson Hirokazu Watanabe, and Mauricio Aredes

82. *Practical Lighting Design with LEDs, Second Edition*
Ron Lenk and Carol Lenk

83. *Introduction to AC Machine Design*
Thomas A. Lipo

84. *Advances in Electric Power and Energy Systems: Load and Price Forecasting*
Mohamed E. El-Hawary

85. *Electricity Markets: Theories and Applications*
Jeremy Lin and Jernando H. Magnago

86. *Multiphysics Simulation by Design for Electrical Machines, Power Electronics and Drives*
Marius Rosu, Ping Zhou, Dingsheng Lin, Dan M. Ionel, Mircea Popescu, Frede Blaabjerg, Vandana Rallabandi, and David Staton

87. *Modular Multilevel Converters: Analysis, Control, and Applications*
Sixing Du, Apparao Dekka, Bin Wu, and Navid Zargari

88. *Electrical Railway Transportation Systems*
Morris Brenna, Federica Foiadelli, and Dario Zaninelli

89. *Energy Processing and Smart Grid*
James A. Momoh

90. *Handbook of Large Turbo-Generator Operation and Maintenance, 3rd Edition*
Geoff Klempner and Isidor Kerszenbaum

91. *Advanced Control of Doubly Fed Induction Generator for Wind Power Systems*
Dehong Xu, Dr. Frede Blaabjerg, Wenjie Chen, and Nan Zhu

92. *Electric Distribution Systems, 2nd Edition*
Abdelhay A. Sallam and Om P. Malik

93. *Power Electronics in Renewable Energy Systems and Smart Grid: Technology and Applications*
Bimal K. Bose

94. *Distributed Fiber Optic Sensing and Dynamic Rating of Power Cables*
Sudhakar Cherukupalli and George J. Anders

95. *Power System and Control and Stability, Third Edition*
Vijay Vittal, James D. McCalley, Paul M. Anderson, and A.A. Fouad

96. *Electromechanical Motion Devices: Rotating Magnetic Field-Based Analysis and Online Animations, Third Edition*
Paul Krause, Oleg Wasynczuk, Steven D. Pekarek, and Timothy O'Connell

97. *Applications of Modern Heuristic Optimization Methods in Power and Energy Systems*
Kwang Y. Lee and Zita A. Vale

98. *Handbook of Large Hydro Generators: Operation and Maintenance*
Glenn Mottershead, Stefano Bomben, Isidor Kerszenbaum, and Geoff Klempner

99. *Advances in Electric Power and Energy: Static State Estimation*
Mohamed E. El-hawary

100. *Arc Flash Hazard Analysis and Mitigation, Second Edition*
J.C. Das

101. *Maintaining Mission Critical Systems in a 24/7 Environment, Third Edition*
Peter M. Curtis

102. *Real-Time Electromagnetic Transient Simulation of AC-DC Networks*
Venkata Dinavahi and Ning Lin

103. *Probabilistic Power System Expansion Planning with Renewable Energy Resources and Energy Storage Systems*
Jaeseok Choi and Kwang Y. Lee

104. *Power Magnetic Devices: A Multi-Objective Design Approach, Second Edition*
Scott D. Sudhoff

105. *Optimal Coordination of Power Protective Devices with Illustrative Examples*
Ali R. Al-Roomi

106. *Resilient Control Architectures and Power Systems*
Craig Rieger, Ronald Boring, Brian Johnson, and Timothy McJunkin

107. *Alternative Liquid Dielectrics for High Voltage Transformer Insulation Systems: Performance Analysis and Applications*
Edited by U. Mohan Rao, I. Fofana, and R. Sarathi

108. *Introduction to the Analysis of Electromechanical Systems*
Paul C. Krause, Oleg Wasynczuk, and Timothy O'Connell

109. *Power Flow Control Solutions for a Modern Grid using SMART Power Flow Controllers*
Kalyan K. Sen and Mey Ling Sen

110. *Power System Protection: Fundamentals and Applications*
John Ciufo and Aaron Cooperberg

111. *Soft-Switching Technology for Three-phase Power Electronics Converters*
Dehong Xu, Rui Li, Ning He, Jinyi Deng, and Yuying Wu

112. *Power System Protection, Second Edition*
Paul M. Anderson, Charles Henville, Rasheek Rifaat, Brian Johnson, and Sakis Meliopoulos

113. *High Voltage and Electrical Insulation Engineering, Second Edition*
Ravindra Arora and Wolfgang Mosch

114. *Modeling and Control of Modern Electrical Energy Systems*
Masoud Karimi-Ghartemani

115. *Control of Power Electronic Converters with Microgrid Applications*
Armdam Ghosh and Firuz Zare

116. *Coordinated Operation and Planning of Modern Heat and Electricity Incorporated Networks*
Mohammadreza Daneshvar, Behnam Mohammadi-Ivatloo, and Kazem Zare

117. *Smart Energy for Transportation and Health in a Smart City*
Chun Sing Lai, Loi Lei Lai, and Qi Hong Lai

118. *Wireless Power Transfer: Principles and Applications*
Zhen Zhang and Hongliang Pang

119. *Intelligent Data Mining and Analysis in Power and Energy Systems: Models and Applications for Smarter Efficient Power Systems*
Zita Vale, Tiago Pinto, Michael Negnevitsky, and Ganesh Kumar Venayagamoorthy

120. *Introduction to Modern Analysis of Electric Machines and Drives*
Paul C. Krause and Thomas C. Krause

121. *Electromagnetic Analysis and Condition Monitoring of Synchronous Generators*
Hossein Ehya and Jawad Faiz

122. *Transportation Electrification: Breakthroughs in Electrified Vehicles, Aircraft, Rolling Stock, and Watercraft*
Ahmed A. Mohamed, Ahmad Arshan Khan, Ahmed T. Elsayed, and Mohamed A. Elshaer

123. *Modular Multilevel Converters: Control, Fault Detection, and Protection*
Fuji Deng, Chengkai Liu, and Zhe Chen

124. *Stability-Constrained Optimization for Modern Power System Operation and Planning*
Yan Xu, Yuan Chi, and Heling Yuan

125. *Interval Methods for Uncertain Power System Analysis*
Alfredo Vaccaro

126. *Practical Partial Discharge Measurement on Electrical Equipment*
Greg Stone, Andrea Cavallini, Glenn Behrmann, and Claudio Angelo Serafino

127. *Graph Database and Graph Computing for Power System Analysis*
Renchang Dai and Guangyi Liu

128. *The Power of Artificial Intelligence for the Next-Generation Oil and Gas Industry: Envisaging AI-inspired Intelligent Energy Systems and Environments*
Pethuru Raj Chelliah, Venkatraman Jayasankar, Mats Agerstam, B. Sundaravadivazhagan, and Robin Cyriac

129. *Microgrids: Theory and Practice*
Peng Zhang

130. *Electric Machinery and Drives: An Electromagnetics Perspective*
Nabeel A. O. Demerdash, JiangBiao He, and Hao Chen

131. *Smart Cyber-Physical Power Systems, Volume 1: Fundamental Concepts, Challenges, and Solutions*
Ali Parizad, Hamid Reza Baghaee, and Saifur Rahman

132. *Smart Cyber-Physical Power Systems, Volume 2: Solutions from Emerging Technologies*
Ali Parizad, Hamid Reza Baghaee, and Saifur Rahman

133. *Analysis of Electric Machinery and Drive Systems, Fourth Edition*
Paul C. Krause, Oleg Wasynczuk, Scott D. Sudhoff, and Steven D. Pekarek

134. *Understanding Electromagnetic Transients in Power System*
Luiz Cera Zanetta Jr.

135. *Communities for Clean Energy Justice and Equity in Grid Modernization*
Mohammadreza Daneshvar, Behnam Mohammadi-Ivatloo, and Amjad Anvari-Moghaddam

136. *Modern Power System*
Arindam Ghosh

137. *Hidden Risk in Occupational Electrical Safety: Through the Lens of System Safety*
H. Landis "Lanny" Floyd

Printed and bound by CPI Group (UK) Ltd, Croydon, CR0 4YY
17/09/2026
14964331-0002